BRIDGE ENGINEERING

Construction and Maintenance

BRIDGE ENGINEERING

Construction and Maintenance

EDITED BY
Wai-Fah Chen
Lian Duan

CRC Press
Taylor & Francis Group
Boca Raton London New York

CRC Press is an imprint of the
Taylor & Francis Group, an **informa** business

The material in this book was previously published in *Bridge Engineering Handbook*, W.F. Chen and L. Duan, Eds., CRC Press, Boca Raton, FL, 2000.

CRC Press
Taylor & Francis Group
6000 Broken Sound Parkway NW, Suite 300
Boca Raton, FL 33487-2742

©2003 by Taylor & Francis Group, LLC
CRC Press is an imprint of Taylor & Francis Group, an Informa business

First issued in paperback 2019

No claim to original U.S. Government works

ISBN-13: 978-0-367-45457-9 (pbk)
ISBN-13: 978-0-8493-1684-5 (hbk)

Visit the Taylor & Francis Web site at
http://www.taylorandfrancis.com

and the CRC Press Web site at
http://www.crcpress.com

Library of Congress Cataloging-in-Publication Data

Catalog record is available from the Library of Congress

Foreword

Among all engineering subjects, bridge engineering is probably the most difficult on which to compose a handbook because it encompasses various fields of arts and sciences. It not only requires knowledge and experience in bridge design and construction, but often involves social, economic, and political activities. Hence, I wish to congratulate the editors and authors for having conceived this thick volume and devoted the time and energy to complete it in such short order. Not only is it the first handbook of bridge engineering as far as I know, but it contains a wealth of information not previously available to bridge engineers. It embraces almost all facets of bridge engineering except the rudimentary analyses and actual field construction of bridge structures, members, and foundations. Of course, bridge engineering is such an immense subject that engineers will always have to go beyond a handbook for additional information and guidance.

I may be somewhat biased in commenting on the background of the two editors, who both came from China, a country rich in the pioneering and design of ancient bridges and just beginning to catch up with the modern world in the science and technology of bridge engineering. It is particularly to the editors' credit to have convinced and gathered so many internationally recognized bridge engineers to contribute chapters. At the same time, younger engineers have introduced new design and construction techniques into the treatise.

This handbook is divided into four volumes, namely:

Superstructure Design
Substructure Design
Seismic Design
Construction and Maintenance

There are 67 chapters, beginning with bridge concepts and aesthestics, two areas only recently emphasized by bridge engineers. Some unusual features, such as rehabilitation, retrofit, and maintenance of bridges, are presented in great detail. The section devoted to seismic design includes soil-foundation-structure interaction. Another section describes and compares bridge engineering practices around the world. I am sure that these special areas will be brought up to date as the future of bridge engineering develops.

I advise each bridge engineer to have a desk copy of this volume with which to survey and examine both the breadth and depth of bridge engineering.

T. Y. Lin
Professor Emeritus, University of California at Berkeley
Chairman, Lin Tung-Yen China, Inc.

Preface

The *Bridge Engineering Handbook* is a unique, comprehensive, and the state-of-the-art reference work and resource book covering the major areas of bridge engineering with the theme *"Bridge to the Twenty-First Century."* It has been written with practicing bridge and structural engineers in mind. The ideal reader will be an M.S.-level structural and bridge engineer with a need for a single reference source to keep abreast of new development and the state-of-the-practice, as well as review standard practices.

The areas of bridge engineering include planning, analysis and design, construction, maintenance and rehabilitation. To provide engineers a well organized and user-friendly, easy to follow resource, the handbook is divided and printed into four Volumes: I Superstructure Design, II Substructure Design, III Seismic Design, and IV Construction and Maintenance.

Volume IV Construction and Maintenance contains constructions of steel and concrete bridges, substructures of major overwater bridges, construction inspections, construction control for cable-stayed bridges, maintenance inspection and rating, strengthening and rehabilitation.

The handbook stresses professional applications and practical solutions. Emphasis has been placed on ready-to-use materials. It contains many formulas and tables that give immediate answers to questions arising from practical works. It describes the basic concepts and assumptions omitting the derivations of formulas and theories. It covers traditional and new, innovative practices. An overview of the structure, organization, and content of the book can be seen by examining the table of contents presented at the beginning of the volume while an in-depth view of a particular subject can be seen by examining the individual table of contents preceding each chapter. References at the end of each chapter can be consulted for more detailed studies.

The chapters have been written by many internationally known authors in different countries covering bridge engineering practices, and research and development in North America, Europe, and Pacific Rim countries. This handbook may provide a glimpse of rapid global economy trend in recent years toward international outsourcing of practice and competition of all dimensions of engineering. In general, the handbook is aimed at the needs of practicing engineers, but materials may be re-organized to accommodate several bridge courses at the undergraduate and graduate levels. The book may also be used as a survey of the practice of bridge engineering around the world.

The authors acknowledge with thanks the comments, suggestions and recommendations during the development of the handbook, by Fritz Leonhardt, Professor Emeritus, Stuttgart University, Germany; Shouji Toma, Professor, Horrai-Gakuen University, Japan; Gerard F. Fox, Consulting Engineer; Jackson L. Durkee, Consulting Engineer; Michael J. Abrahams, Senior Vice President, Parsons Brinckerhoff Quade & Douglas, Inc.; Ben C. Gerwick, Jr., Professor Emeritus, University of California at Berkeley; Gregory F. Fenves, Professor, University of California at Berkeley; John M. Kulicki, President and Chief Engineer, Modjeski and Masters; James Chai, Supervising Transportation Engineer, California Department of Transportation; Jinrong Wang, Senior Bridge Engineer, California Department of Transportation; David W. Liu, Principal, Imbsen & Associates, Inc.

Wai-Fah Chen
Lian Duan

Editors

Wai-Fah Chen is presently Dean of the College of Engineering at the University of Hawaii. He was a George E. Goodwin Distinguished Professor of Civil Engineering and Head of the Department of Structural Engineering at Purdue University from 1976 to 1999.

He received his B.S. in civil engineering from the National Cheng-Kung University, Taiwan, in 1959, M.S. in structural engineering from Lehigh University, Pennsylvania, in 1963, and Ph.D. in solid mechanics from Brown University, Rhode Island, in 1966. He received the Distinguished Alumnus Award from the National Cheng-Kung University in 1988 and the Distinguished Engineering Alumnus Medal from Brown University in 1999.

Dr. Chen's research interests cover several areas, including constitutive modeling of engineering materials, soil and concrete plasticity, structural connections, and structural stability. He is the recipient of several national engineering awards, including the Raymond Reese Research Prize and the Shortridge Hardesty Award, both from the American Society of Civil Engineers, and the T. R. Higgins Lectureship Award from the American Institute of Steel Construction. In 1995, he was elected to the U.S. National Academy of Engineering. In 1997, he was awarded Honorary Membership by the American Society of Civil Engineers. In 1998, he was elected to the Academia Sinica (National Academy of Science) in Taiwan.

A widely respected author, Dr. Chen authored and coauthored more than 20 engineering books and 500 technical papers. His books include several classical works such as *Limit Analysis and Soil Plasticity* (Elsevier, 1975), the two-volume *Theory of Beam-Columns* (McGraw-Hill, 1976–77), *Plasticity in Reinforced Concrete* (McGraw-Hill, 1982), and the two-volume *Constitutive Equations for Engineering Materials* (Elsevier, 1994). He currently serves on the editorial boards of more than 10 technical journals. He has been listed in more than 20 *Who's Who* publications.

Dr. Chen is the editor-in-chief for the popular 1995 *Civil Engineering Handbook*, the 1997 *Handbook of Structural Engineering*, and the 1999 *Bridge Engineering Handbook*. He currently serves as the consulting editor for McGraw-Hill's *Encyclopedia of Science and Technology*.

He has been a longtime member of the Executive Committee of the Structural Stability Research Council and the Specification Committee of the American Institute of Steel Construction. He has been a consultant for Exxon Production Research on offshore structures, for Skidmore, Owings, and Merrill in Chicago on tall steel buildings, and for the World Bank on the Chinese University Development Projects, among many others.

Dr. Chen has taught at Lehigh University, Purdue University, and the University of Hawaii.

Lian Duan is a Senior Bridge Engineer with the California Department of Transportation (Caltrans), U.S., and Professor of Structural Engineering at Taiyuan University of Technology, China. He received his B.S. in civil engineering in 1975, M.S. in structural engineering in 1981 from Taiyuan University of Technology, and Ph.D. in structural engineering from Purdue University, West Lafayette, Indiana, in 1990. Dr. Duan worked at the Northeastern China Power Design Institute from 1975 to 1978.

Dr. Duan's research interests cover areas including inelastic behavior of reinforced concrete and steel structures, structural stability, and seismic bridge analysis and design. With more than 60 authored or coauthored papers, chapters, and reports, his research focuses on the development of unified interaction equations for steel beam-columns, flexural stiffness of reinforced concrete members, effective length factors of compression members, and design of bridge structures.

Dr. Duan is also an esteemed practicing engineer and a registered P.E. in California. He has designed numerous building and bridge structures. He was lead engineer for the development of the seismic retrofit design criteria for the San Francisco-Oakland Bay Bridge West Spans and made significant contributions to the project. He is co-editor of the *Structural Engineering Handbook* CRCnetBase 2000 (CRC Press, 2000) and the *Bridge Engineering Handbook* (CRC Press, 2000), winner of *Choice Magazine's* Outstanding Academic Title Award for 2000. He received the ASCE 2001 Arthur M. Wellington Prize for his paper "Section Properties for Latticed Members of San Francisco–Oakland Bay Bridge." He currently serves as Caltrans Structural Steel Committee Chairman and a member of the Transportation Research Board A2C02 Steel Bridge Committee.

Contributors

Masoud Alemi
California Department of
    Transportation
Sacramento, California

Michael M. Blank
U.S. Army Corps of Engineers
Philadelphia, Pennsylvania

Simon A. Blank
California Department of
    Transportation
Walnut Creek, California

Jackson Durkee
Consulting Structural Engineer
Bethlehem, Pennsylvania

Mahmoud Fustok
California Department of
    Transportation
Sacramento, California

Ben C. Gerwick, Jr.
Ben C. Gerwick, Inc.
Consulting Engineer
San Francisco, California

Danjian Han
Department of Civil Engineering
South China University of
    Technology
Guangzhou, China

F. Wayne Klaiber
Department of Civil Engineering
Iowa State University
Ames, Iowa

Luis R. Luberas
U.S. Army Corps of Engineers
Philadelphia, Pennsylvania

**Murugesu
Vinayagamoorthy**
California Department of
    Transportation
Sacramento, California

Terry J. Wipf
Department of Civil Engineering
Iowa State University
Ames, Iowa

Quansheng Yan
College of Traffic and
    Communication
South China University of
    Technology
Guangzhou, China

Contents

1

Steel Bridge Construction

Jackson Durkee
Consulting Structural Engineer,
Bethlehem, Pa.

0-8493-1684-7/03/$0.00+$1.50
© 2003 by CRC Press LLC

1.1 Introduction

This chapter addresses some of the principles and practices applicable to the construction of medium- and long-span steel bridges — structures of such size and complexity that construction engineering becomes an important or even the governing factor in the successful fabrication and erection of the superstructure steelwork.

We begin with an explanation of the fundamental nature of construction engineering, then go on to explain some of the challenges and obstacles involved. The basic considerations of cambering are explained. Two general approaches to the fabrication and erection of bridge steelwork are described, with examples from experience with arch bridges, suspension bridges, and cable-stayed bridges.

The problem of erection-strength adequacy of trusswork under erection is considered, and a method of appraisal offered that is believed to be superior to the standard working-stress procedure.

Typical problems with respect to construction procedure drawings, specifications, and practices are reviewed, and methods for improvement suggested. The need for comprehensive bridge erection-engineering specifications, and for standard conditions for contracting, is set forth, and the design-and-construct contracting procedure is described.

Finally, we take a view ahead, to the future prospects for effective construction engineering in the U.S.

The chapter also contains a large number of illustrations showing a variety of erection methods for several types of major steel bridges.

1.2 Construction Engineering in Relation to Design Engineering

With respect to bridge steelwork the differences between construction engineering and design engineering should be kept firmly in mind. Design engineering is of course a concept and process well known to structural engineers; it involves preparing a set of plans and specifications — known as the contract documents — that define the structure in its completed configuration, referred to as the geometric outline. Thus, the design drawings describe to the contractor the steel bridge superstructure that the owner wants to see in place when the project is completed. A considerable design engineering effort is required to prepare a good set of contract documents.

Construction engineering, however, is not so well known. It involves governing and guiding the fabrication and erection operations needed to produce the structural steel members to the proper cambered or "no-load" shape, and get them safely and efficiently "up in the air" in place in the structure, so that the completed structure under the deadload conditions and at normal temperature will meet the geometric and stress requirements stipulated on the design drawings.

Four key considerations may be noted: (1) design engineering is widely practiced and reasonably well understood, and is the subject of a steady stream of technical papers; (2) construction engineering is practiced on only a limited basis, is not as well understood, and is hardly ever discussed; (3) for medium- and long-span bridges, the construction engineering aspects are likely to be no less important than design engineering aspects; and (4) adequately staffed and experienced construction-engineering offices are a rarity.

1.3 Construction Engineering Can Be Critical

The construction phase of the total life of a major steel bridge will probably be much more hazardous than the service-use phase. Experience shows that a large bridge is more likely to suffer failure during erection than after completion. Many decades ago, steel bridge design engineering had progressed to the stage where the chance of structural failure under service loadings became altogether remote. However, the erection phase for a large bridge is inherently less secure, primarily because of the prospect of inadequacies in construction engineering and its implementation at the job site. The hazards associated with the erection of large steel bridges will be readily apparent from a review of the illustrations in this chapter.

For significant steel bridges the key to construction integrity lies in the proper planning and engineering of steelwork fabrication and erection. Conversely, failure to attend properly to construction engineering

FIGURE 1.1　Failure of a steel girder bridge during erection, 1995. Steel bridge failures such as this one invite suspicion that the construction engineering aspects were not properly attended to.

constitutes an invitation to disaster. In fact, this thesis is so compelling that whenever a steel bridge failure occurs during construction (see for example Figure 1.1), it is reasonable to assume that the construction engineering investigation was either inadequate, not properly implemented, or both.

1.4　Premises and Objectives of Construction Engineering

During the erection sequences the various components of steel bridges may be subjected to stresses that are quite different from those which will occur under the service loadings and which have been provided for by the designer. For example, during construction there may be a derrick moving and working on the partially erected structure, and the structure may be cantilevered out some distance causing tension-designed members to be in compression and vice versa. Thus, the steelwork contractor needs to engineer the bridge members through their various construction loadings, and strengthen and stabilize them as may be necessary. Further, the contractor may need to provide temporary members to support and stabilize the structure as it passes through its successive erection configurations.

In addition to strength problems there are also geometric considerations. The steelwork contractor must engineer the construction sequences step-by-step to ensure that the structure will fit properly together as erection progresses, and that the final or closing members can be moved into position and connected. Finally, of course, the steelwork contractor must carry out the engineering studies needed to ensure that the geometry and stressing of the completed structure under normal temperature will be in accordance with the requirements of the design plans and specifications.

1.5　Fabrication and Erection Information Shown on Design Plans

Regrettably, the level of engineering effort required to accomplish safe and efficient fabrication and erection of steelwork superstructures is not widely understood or appreciated in bridge design offices, nor indeed by many steelwork contractors. It is only infrequently that we find a proper level of capability and effort in the engineering of construction.

The design drawings for an important bridge will sometimes display an erection scheme, even though most designers are not experienced in the practice of erection engineering and usually expend only a minimum or even superficial effort on erection studies. The scheme portrayed may not be practical, or may not be suitable in respect to the bidder or contractor's equipment and experience. Accordingly, the bidder or contractor may be making a serious mistake if he relies on an erection scheme portrayed on the design plans.

As an example of misplaced erection effort on the part of the designer, there have been cases where the design plans show cantilever erection by deck travelers, with the permanent members strengthened correspondingly to accommodate the erection loadings; but the successful bidder elected to use waterborne erection derricks with long booms, thereby obviating the necessity for most or all of the erection strengthening provided on the design plans. Further, even in those cases where the contractor would decide to erect by cantilevering as anticipated on the plans, there is hardly any way for the design engineer to know what will be the weight and dimensions of the contractor's erection travelers.

1.6   Erection Feasibility

Of course, the bridge designer does have a certain responsibility to his client and to the public in respect to the erection of the bridge steelwork. This responsibility includes: (1) making certain, during the design stage, that there is a feasible and economical method to erect the steelwork; (2) setting forth in the contract documents any necessary erection guidelines and restrictions; and (3) reviewing the contractor's erection scheme, including any strengthening that may be needed, to verify its suitability. It may be noted that this latter review does not relieve the contractor from responsibility for the adequacy and safety of the field operations.

Bridge annals include a number of cases where the design engineer failed to consider erection feasibility. In one notable instance the design plans showed the 1200 ft (366 m) main span for a long crossing over a wide river as an esthetically pleasing steel tied-arch. However, erection of such a span in the middle of the river was impractical; one bidder found that the tonnage of falsework required was about the same as the weight of the permanent arch-span steelwork. Following opening of the bids, the owner found the prices quoted to be well beyond the resources available, and the tied-arch main span was discarded in favor of a through-cantilever structure, for which erection falsework needs were minimal and practical.

It may be noted that design engineers can stand clear of serious mistakes such as this one, by the simple expedient of conferring with prospective bidders during the preliminary design stage of a major bridge.

1.7   Illustrations of Challenges in Construction Engineering

Space does not permit comprehensive coverage of the numerous and difficult technical challenges that can confront the construction engineer in the course of the erection of various types of major steel bridges. However, some conception of the kinds of steelwork erection problems, the methods available to resolve them, and the hazards involved can be conveyed by views of bridges in various stages of erection; refer to the illustrations in the text.

1.8   Obstacles to Effective Construction Engineering

There is an unfortunate tendency among design engineers to view construction engineering as relatively unimportant. This view may be augmented by the fact that few designers have had any significant experience in the engineering of construction.

Further, managers in the construction industry must look critically at costs, and they can readily develop the attitude that their engineers are doing unnecessary theoretical studies and calculations, detached from the practical world. (And indeed, this may sometimes be the case.) Such management apprehension can constitute a serious obstacle to staff engineers who see the need to have enough money in the bridge tender to cover a proper construction engineering effort for the project. There is the tendency for steelwork construction company management to cut back the construction engineering allowance, partly because of this apprehension and partly because of the concern that other tenderers will not be allotting adequate money for construction engineering. This effort is often thought of by company management as "a necessary evil" at best — something they would prefer not to be bothered with or burdened with.

Accordingly, construction engineering tends to be a difficult area of endeavor. The way for staff engineers to gain the confidence of management is obvious — they need to conduct their investigations to a level of technical proficiency that will command management respect and support, and they must keep management informed as to what they are doing and why it is necessary. As for management's concern that other bridge tenderers will not be putting into their packages much money for construction engineering, this concern is no doubt often justified, and it is difficult to see how responsible steelwork contractors can cope with this problem.

1.9 Examples of Inadequate Construction Engineering Allowances and Effort

Even with the best of intentions, the bidder's allocation of money to construction engineering can be inadequate. A case in point involved a very heavy, long-span cantilever truss bridge crossing a major river. The bridge superstructure carried a contract price of some $30 million, including an allowance of $150,000, or about one-half of 1%, for construction engineering of the permanent steelwork (i.e., not including such matters as design of erection equipment). As fabrication and erection progressed, many unanticipated technical problems came forward, including brittle-fracture aspects of certain grades of the high-strength structural steel, and aerodynamic instability of H-shaped vertical and diagonal truss members. In the end the contractor's construction engineering effort mounted to about $1.3 million, almost nine times the estimated cost.

Another significant example — this one in the domain of buildings — involved a design-and-construct project for airplane maintenance hangars at a prominent international airport. There were two large and complicated buildings, each 100×150 m (328×492 ft) in plan and 37 m (121 ft) high with a 10 m (33 ft) deep space-frame roof. Each building contained about 2450 tons of structural steelwork. The design-and-construct steelwork contractor had submitted a bid of about $30 million, and included therein was the magnificent sum of $5,000 for construction engineering, under the expectation that this work could be done on an incidental basis by the project engineer in his "spare time."

As the steelwork contract went forward it quickly became obvious that the construction engineering effort had been grossly underestimated. The contractor proceeded to staff-up appropriately and carried out in-depth studies, leading to a detailed erection procedure manual of some 270 pages showing such matters as erection equipment and its positioning and clearances; falsework requirements; lifting tackle and jacking facilities; stress, stability, and geometric studies for gravity and wind loads; step-by-step instructions for raising, entering, and connecting the steelwork components; closing and swinging the roof structure and portal frame; and welding guidelines and procedures. This erection procedure manual turned out to be a key factor in the success of the fieldwork. The cost of this construction engineering effort amounted to about ten times the estimate, but still came to a mere one-fifth of 1% of the total contract cost.

In yet another example a major steelwork general contractor was induced to sublet the erection of a long-span cantilever truss bridge to a reputable erection contractor, whose quoted *price* for the work was less than the general contractor's estimated *cost*. During the erection cycle the general contractor's engineers made some visits to the job site to observe progress, and were surprised and disconcerted to observe how little erection engineering and planning had been accomplished. For example, the erector had made no provision for installing jacks in the bottom-chord jacking points for closure of the main span; it was left up to the field forces to provide the jack bearing components inside the bottom-chord joints and to find the required jacks in the local market. When the job-built installations were tested it was discovered that they would not lift the cantilevered weight, and the job had to be shut down while the field engineer scouted around to find larger-capacity jacks. Further, certain compression members did not appear to be properly braced to carry the erection loadings; the erector had not engineered those members, but just assumed they were adequate. It became obvious that the erector had not appraised the bridge members for erection adequacy and had done little or no planning and engineering of the critical evolutions to be carried out in the field.

Many further examples of inadequate attention to construction engineering could be presented. Experience shows that the amounts of money and time allocated by steelwork contractors for the engineering of construction are frequently far less than desirable or necessary. Clearly, effort spent on construction engineering is worthwhile; it is obviously more efficient and cheaper, and certainly much safer, to plan and engineer steelwork construction in the office in advance of the work, rather than to leave these important matters for the field forces to work out. Just a few bad moves on site, with the corresponding waste of labor and equipment hours, will quickly use up sums of money much greater than those required for a proper construction engineering effort — not to mention the costs of any job accidents that might occur.

The obvious question is "Why is construction engineering not properly attended to?" Do not contractors learn, after a bad experience or two, that it is both necessary and cost effective to do a thorough job of planning and engineering the construction of important bridge projects? Experience and observation would seem to indicate that some steelwork contractors learn this lesson, while many do not. There is always pressure to reduce bid prices to the absolute minimum, and to add even a modest sum for construction engineering must inevitably reduce the prospect of being the low bidder.

1.10   Considerations Governing Construction Engineering Practices

There are no textbooks or manuals that define how to accomplish a proper job of construction engineering. In bridge construction (and no doubt in building construction as well) the engineering of construction tends to be a matter of each firm's experience, expertise, policies, and practices. Usually there is more than one way to build the structure, depending on the contractor's ingenuity and engineering skill, his risk appraisal and inclination to assume risk, the experience of his fabrication and erection work forces, his available equipment, and his personal preferences. Experience shows that each project is different; and although there will be similarities from one bridge of a given type to another, the construction engineering must be accomplished on an individual project basis. Many aspects of the project at hand will turn out to be different from those of previous similar jobs, and also there may be new engineering considerations and requirements for a given project that did not come forward on previous similar work.

During the estimating and bidding phase of the project the prudent, experienced bridge steelwork contractor will "start from scratch" and perform his own fabrication and erection studies, irrespective of any erection schemes and information that may be shown on the design plans. These studies can involve a considerable expenditure of both time and money, and thereby place that contractor at a disadvantage in respect to those bidders who are willing to rely on hasty, superficial studies, or — where the design engineer has shown an erection scheme — to simply assume that it has been engineered correctly and proceed to use it. The responsible contractor, on the other hand, will appraise the feasible construction methods and evaluate their costs and risks, and then make his selection.

After the contract has been executed the contractor will set forth how he intends to fabricate and erect, in detailed plans that could involve a large number of calculation sheets and drawings along with construction procedure documents. It is appropriate for the design engineer on behalf of his client to review the contractor's plans carefully, perform a check of construction considerations, and raise appropriate questions. Where the contractor does not agree with the designer's comments the two parties get together for review and discussion, and in the end they concur on essential factors such as fabrication and erection procedures and sequences, the weight and positioning of erection equipment, the design of falsework and other temporary components, erection stressing and strengthening of the permanent steelwork, erection stability and bracing of critical components, any erection check measurements that may be needed, and span closing and swinging operations.

The design engineer's approval is needed for certain fabrication plans, such as the cambering of individual members; however, in most cases the designer should stand clear of actual *approval* of the contractor's construction plans since he is not in a position to accept construction responsibility, and too many things can happen during the field evolutions over which the designer has no control.

It should be emphasized that even though the design engineer has usually had no significant experience in steelwork construction, the contractor should welcome his comments and evaluate them carefully and respectfully. In major bridge projects many construction matters can be improved upon or get out of control, and the contractor should take advantage of every opportunity to augment his prospects and performance. The experienced contractor will make sure that he works constructively with the design engineer, standing well clear of antagonistic or confrontational posturing.

1.11   Camber Considerations

One of the first construction engineering problems to be resolved by the steel bridge contractor is the cambering of individual bridge components. The design plans will show the "geometric outline" of the bridge, which is its shape under the designated load condition — commonly full dead load — at normal temperature. The contractor, however, fabricates the bridge members under the no-load condition, and at the "shop temperature" — the temperature at which the shop measuring tapes have been standardized and will have the correct length. The difference between the shape of a member under full dead load and normal temperature, and its shape at the no-load condition and shop temperature, is defined as member camber.

While camber is inherently a simple concept, it is frequently misunderstood; indeed, it is often not correctly defined in design specifications and contract documents. For example, beam and girder camber has been defined in specifications as "the convexity induced into a member to provide for vertical curvature of grade and to offset the anticipated deflections indicated on the plans when the member is in its erected position in the structure. Cambers shall be measured in this erected position." This definition is not correct, and reflects a common misunderstanding of a key structural engineering term. Camber of bending members is not convexity, nor does it have anything to do with grade vertical curvature, nor is it measured with the member in the erected position. Camber — of a bending member, or any other member — is the *difference in shape* of the member under its no-load fabrication outline as compared with its geometric outline; and it is "measured" — i.e., the cambered dimensions are applied to the member — not when it is in the *erected* position (whatever that might be), but rather, when it is in the *no-load* condition.

In summary, camber is a *difference* in shape and not the shape itself. Beams and girders are commonly cambered to compensate for deadload bending, and truss members to compensate for deadload axial force. However, further refinements can be introduced as may be needed; for example, the arch-rib box members of the Lewiston-Queenston bridge (Figure 1.4) were cambered to compensate for deadload axial force, bending, and shear.

A further common misunderstanding regarding cambering of bridge members involves the effect of the erection scheme on cambers. The erection scheme may require certain members to be strengthened, and this in turn will affect the cambers of those members (and possibly of others as well, in the case of statically indeterminate structures). However, the fabricator should address the matter of cambering only after the final sizes of all bridge members have been determined. Camber is a function of member properties, and there is no merit to calculating camber for members whose cross-sectional areas may subsequently be increased because of erection forces.

Thus, the erection scheme may affect the required member properties, and these in turn will affect member cambering; but the erection scheme does not *of itself* have any effect on camber. Obviously, the temporary stress-and-strain maneuvers to which a member will be subjected, between its no-load condition in the shop and its full-deadload condition in the completed structure, can have no bearing on the camber calculations for the member.

To illustrate the general principles that govern the cambering procedure, consider the main trusses of a truss bridge. The first step is to determine the erection procedure to be used, and to augment the strength of the truss members as may be necessary to sustain the erection forces. Next, the bridge deadload weights are determined, and the member deadload forces and effective cross-sectional areas are calculated.

Consider now a truss chord member having a geometric length of 49.1921 ft panel-point-to-panel-point and an effective cross-sectional area of 344.5 in.^{2}, carrying a deadload compressive force of 4230 kips. The bridge normal temperature is 45°F and the shop temperature is 68°F. We proceed as follows:

1. Assume that the chord member is in place in the bridge, at the full dead load of –4230 kips and the normal temperature of 45°F.
2. Remove the member from the bridge, allowing its compressive force to fall to zero. The member will increase in length by an amount ΔL_s:

$$\Delta L_s = \frac{SL}{AE} = \frac{4230\ kips \times 49.1921\ ft}{344.5\ in.^2 \times 29000\ kips/in.^2}$$

$$= 0.0208\,ft$$

3. Now raise the member temperature from 45°F to 68°F. The member will increase in length by an additional amount ΔL_t:

$$\Delta L_t = L\omega t = (49.1921 + 0.0208)\ ft \times$$

$$0.0000065/\deg \times (68-45)\deg$$

$$= 0.0074\ ft$$

4. The total increase in member length will be:

$$\Delta L = \Delta L_s + \Delta L_t = 0.0208 + 0.0074$$

$$= 0.0282\ ft$$

5. The theoretical cambered member length — the no-load length at 68°F — will be:

$$L_{tc} = 49.1921 + 0.0282 = 49.2203\ ft$$

6. Rounding L_{tc} to the nearest 1/32 in., we obtain the cambered member length for fabrication as:

$$L_{fc} = 49\ ft\ 2\frac{21}{32}\ in.$$

Accordingly, the general procedure for cambering a bridge member of any type can be summarized as follows:

1. Strengthen the structure to accommodate erection forces, as may be needed.
2. Determine the bridge deadload weights, and the corresponding member deadload forces and effective cross-sectional areas.
3. Starting with the structure in its geometric outline, remove the member to be cambered.
4. Allow the deadload force in the member to fall to zero, thereby changing its shape to that corresponding to the no-load condition.

5. Further change the shape of the member to correspond to that at the shop temperature.
6. Accomplish any rounding of member dimensions that may be needed for practical purposes.
7. The total change of shape of the member — from geometric (at normal temperature) to no-load at shop temperature — constitutes the member camber.

It should be noted that the gusset plates for bridge-truss joints are always fabricated with the connecting-member axes coming in at their *geometric* angles. As the members are erected and the joints fitted-up, secondary bending moments will be induced at the truss joints under the steel-load-only condition; but these secondary moments will disappear when the bridge reaches its full-deadload condition.

1.12 Two General Approaches to Fabrication and Erection of Bridge Steelwork

As has been stated previously, the objective in steel bridge construction is to fabricate and erect the structure so that it will have the geometry and stressing designated on the design plans, under full dead-load at normal temperature. This geometry is known as the geometric outline. In the case of steel bridges there have been, over the decades, two general procedures for achieving this objective:

1. The "field adjustment" procedure — Carry out a continuing program of steelwork surveys and measurements in the field as erection progresses, in an attempt to discover fabrication and erection deficiencies; and perform continuing steelwork adjustments in an effort to compensate for such deficiencies and for errors in span baselines and pier elevations.
2. The "shop control" procedure — Place total reliance on first-order surveying of span baselines and pier elevations, and on accurate steelwork fabrication and erection augmented by meticulous construction engineering; and proceed with erection without any field adjustments, on the basis that the resulting bridge deadload geometry and stressing will be as good as can possibly be achieved.

Bridge designers have a strong tendency to overestimate the capability of field forces to accomplish accurate measurements and effective adjustments of the partially erected structure, and at the same time they tend to underestimate the positive effects of precise steel bridgework fabrication and erection. As a result, we continue to find contract drawings for major steel bridges that call for field evolutions such as the following:

1. **Continuous trusses and girders** — At the designated stages, measure or "weigh" the reactions on each pier, compare them with calculated theoretical values, and add or remove bearing-shoe shims to bring measured values into agreement with calculated values.
2. **Arch bridges** — With the arch ribs erected to midspan and only the short, closing "crown sections" not yet in place, measure thrust and moment at the crown, compare them with calculated theoretical values, and then adjust the shape of the closing sections to correct for errors in span-length measurements and in bearing-surface angles at skewback supports, along with accumulated fabrication and erection errors.
3. **Suspension bridges** — Following erection of the first cable wire or strand across the spans from anchorage to anchorage, survey its sag in each span and adjust these sags to agree with calculated theoretical values.
4. **Arch bridges and suspension bridges** — Carry out a deck-profile survey along each side of the bridge under the steel-load-only condition, compare survey results with the theoretical profile, and shim the suspender sockets so as to render the bridge floorbeams level in the completed structure.
5. **Cable-stayed bridges** — At each deck-steelwork erection stage, adjust tensions in the newly erected cable stays so as to bring the surveyed deck profile and measured stay tensions into agreement with calculated theoretical data.

FIGURE 1.2 Erection of arch ribs, Rainbow Bridge, Niagara Falls, New York, 1941. Bridge span is 950 ft (290 m), with rise of 150 ft (46 m); box ribs are 3 × 12 ft (0.91 × 3.66 m). Tiebacks were attached starting at the end of the third tier and jumped forward as erection progressed (see Figure 1.3). Much permanent steelwork was used in tieback bents. Derricks on approaches load steelwork onto material cars that travel up arch ribs. Travelers are shown erecting last full-length arch-rib sections, leaving only the short, closing crown sections to be erected. Canada is at right, the U.S. at left. (Courtesy of Bethlehem Steel Corporation.)

There are two prime obstacles to the success of "field adjustment" procedures of whatever type: (1) field determination of the actual geometric and stress conditions of the partially erected structure and its components will not necessarily be definitive, and (2) calculation of the corresponding "proper" or "target" theoretical geometric and stress conditions will most likely prove to be less than authoritative.

1.13 Example of Arch Bridge Construction

In the case of the arch bridge closing sections referred to heretofore, experience on the construction of two major fixed-arch bridges crossing the Niagara River gorge from the U.S. to Canada — the Rainbow and the Lewiston-Queenston arch bridges (see Figures 1.2 through 1.5) — has demonstrated the difficulty, and indeed the futility, of attempts to make field-measured geometric and stress conditions agree with calculated theoretical values. The broad intent for both structures was to make such adjustments in the shape of the arch-rib closing sections at the crown (which were nominally about 1 ft [0.3 m] long) as would bring the arch-rib actual crown moments and thrusts into agreement with the calculated theoretical values, thereby correcting for errors in span-length measurements, errors in bearing-surface angles at the skewback supports, and errors in fabrication and erection of the arch-rib sections.

Following extensive theoretical investigations and on-site measurements the steelwork contractor found, in the case of each Niagara arch bridge, that there were large percentage differences between the field-measured and the calculated theoretical values of arch-rib thrust, moment, and line-of-thrust position, and that the measurements could not be interpreted so as to indicate what corrections to the theoretical closing crown sections, if any, should be made. Accordingly, the contractor concluded that the best solution in each case was to abandon any attempts at correction and simply install the theoretical-shape closing crown sections. In each case, the contractor's recommendation was accepted by the design engineer.

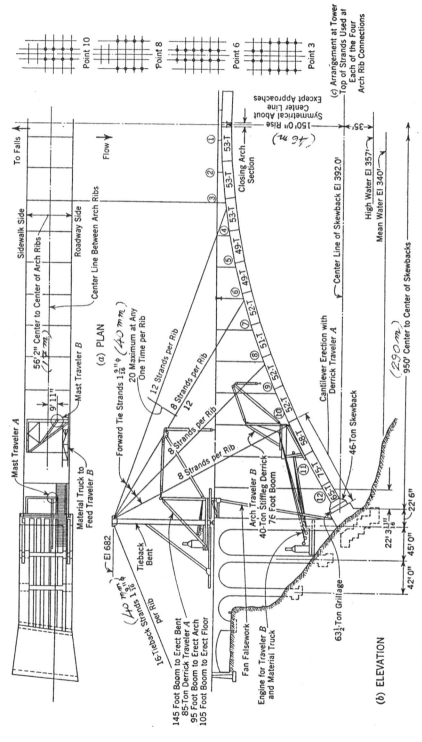

FIGURE 1.3 Rainbow Bridge, Niagara Falls, New York, showing successive arch tieback positions. Arch-rib erection geometry and stressing were controlled by means of measured tieback tensions in combination with surveyed arch-rib elevations.

FIGURE 1.4 Lewiston-Queenston arch bridge, near Niagara Falls, New York, 1962. The longest fixed-arch span in the U.S. at 1000 ft (305 m); rise is 159 ft (48 m). Box arch-rib sections are typically about 3 × 13-1/2 ft (0.9 × 4.1 m) in cross-section and about 44-1/2 ft (13.6 m) long. Job was estimated using erection tiebacks (same as shown in Figure 1.3), but subsequent studies showed the long, sloping falsework bents to be more economical (even if less secure looking). Much permanent steelwork was used in falsework bents. Derricks on approaches load steelwork onto material cars that travel up arch ribs. The 115-ton-capacity travelers are shown erecting the last full-length arch-rib sections, leaving only the short, closing crown sections to be erected. Canada is at left, the U.S. at right. (Courtesy of Bethlehem Steel Corporation.)

Points to be noted in respect to these field-closure evolutions for the two long-span arch bridges are that accurate jack-load closure measurements at the crown are difficult to obtain under field conditions; and calculation of corresponding theoretical crown thrusts and moments are likely to be questionable because of uncertainties in the dead loading, in the weights of erection equipment, and in the steelwork temperature. Therefore, attempts to adjust the shape of the closing crown sections so as to bring the actual stress condition of the arch ribs closer to the presumed theoretical condition are not likely to be either practical or successful.

It was concluded that for long, flexible arch ribs, the best construction philosophy and practice is (1) to achieve overall geometric control of the structure by performing all field survey work and steelwork fabrication and erection operations to a meticulous degree of accuracy, and then (2) to rely on that overall geometric control to produce a finished structure having the desired stressing and geometry. For the Rainbow arch bridge, these practical construction considerations were set forth definitively by the contractor in [2]. The contractor's experience for the Lewiston-Queenston arch bridge was similar to that on Rainbow, and was reported — although in considerably less detail — in [10].

1.14   Which Construction Procedure Is To Be Preferred?

The contractor's experience on the construction of the two long-span fixed-arch bridges is set forth at length since it illustrates a key construction theorem that is broadly applicable to the fabrication and erection of steel bridges of all types. This theorem holds that the contractor's best procedure for achieving, in the completed structure, the deadload geometry and stressing stipulated on the design plans, is generally as follows:

1. Determine deadload stress data for the structure at its geometric outline (under normal temperature), based on accurately calculated weights for all components.
2. Determine the cambered (i.e., "no-load") dimensions of each component. This involves determining the change of shape of each component from the deadload geometry, as its deadload

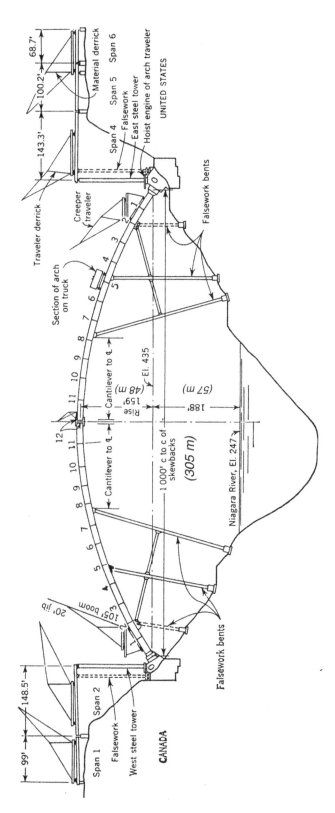

FIGURE 1.5 Lewison-Queenston arch bridge near Niagara Falls, New York. Crawler cranes erect steelwork for spans 1 and 6, and erect material derricks thereon. These derricks erect traveler derricks, which move forward and erect supporting falsework and spans 2, 5, and 4. Traveler derricks erect arch-rib sections 1 and 2 supporting falsework at each skewback, then set up creeper derricks, which erect arches to midspan.

stressing is removed and its temperature is changed from normal to the shop temperature. (Refer to Section 1.11.).

3. Fabricate, with all due precision, each structural component to its proper no-load dimensions — except for certain flexible components such as wire rope and strand members, which may require special treatment.

4. Accomplish shop assembly of members and "reaming assembled" of holes in joints, as needed.

5. Carry out comprehensive engineering studies of the structure under erection at each key erection stage, determining corresponding stress and geometric data, and prepare a step-by-step erection procedure plan, incorporating any check measurements that may be necessary or desirable.

6. During the erection program, bring all members and joints to the designated alignment prior to bolting or welding.

7. Enter and connect the final or closing structural components, following the closing procedure plan, without attempting any field measurements thereof or adjustments thereto.

In summary, the key to construction success is to accomplish the field surveys of critical baselines and support elevations with all due precision, perform construction engineering studies comprehensively and shop fabrication accurately, and then carry the erection evolutions through in the field without any second guessing and ill-advised attempts at measurement and adjustment.

It may be noted that no special treatment is accorded to statically indeterminate members; they are fabricated and erected under the same governing considerations applicable to statically determinate members, as set forth above. It may be noted further that this general steel bridge construction philosophy does not rule out check measurements altogether, as erection goes forward; under certain special conditions, measurements of stressing and/or geometry at critical erection stages may be necessary or desirable in order to confirm structural integrity. However, before the erector calls for any such measurements he should make certain that they will prove to be practical and meaningful.

1.15   Example of Suspension Bridge Cable Construction

In order to illustrate the "shop control" construction philosophy further, its application to the main cables of the first Wm. Preston Lane, Jr., Memorial Bridge, crossing the Chesapeake Bay in Maryland, completed in 1952 (Figure 1.6), will be described. Suspension bridge cables constitute one of the most difficult bridge erection challenges. Up until "first Chesapeake" the cables of major suspension bridges had been adjusted to the correct position in each span by means of a sag survey of the first-erected cable wires or strands, using surveying instruments and target rods. However, on first Chesapeake, with its 1600 ft (488 m) main

FIGURE 1.6 Suspension spans of first Chesapeake Bay Bridge, Maryland, 1952. Deck steelwork is under erection and is about 50% complete. A typical four-panel through-truss deck section, weighing about 100 tons, is being picked in west side span, and also in east side span in distance. Main span is 1600 ft (488 m) and side spans are 661 ft (201 m); towers are 324 ft (99 m) high. Cables are 14 in. (356 mm) in diameter and are made up of 61 helical bridge strands each (see Figure 1.8).

span, 661 ft (201 m) side spans, and 450 ft (137 m) back spans, the steelwork contractor recommended abandoning the standard cable-sag survey and adopting the "setting-to-mark" procedure for positioning the guide strands — a significant new concept in suspension bridge cable construction.

The steelwork contractor's rationale for "setting to marks" was spelled out in a letter to the design engineer (see Figure 1.7). (The complete letter is reproduced because it spells out significant construction philosophies.) This innovation was accepted by the design engineer. It should be noted that the contractor's major argument was that setting to marks would lead to more accurate cable placement than would a sag survey. The minor arguments, alluded to in the letter, were the resulting savings in preparatory office engineering work and in the field engineering effort, and most likely in construction time as well.

Each cable consisted of 61 standard helical-type bridge strands, as shown in Figure 1.8. To implement the setting-to-mark procedure each of three bottom-layer "guide strands" of each cable (i.e., strands 1, 2, and 3) was accurately measured in the manufacturing shop under the simulated full-deadload tension, and circumferential marks were placed at the four center-of-saddle positions of each strand. Then, in the field, the guide strands (each about 3955 ft [1205 m] long) were erected and positioned according to the following procedure:

1. Place the three guide strands for each cable "on the mark" at each of the four saddles and set normal shims at each of the two anchorages.
2. Under conditions of uniform temperature and no wind, measure the sag differences among the three guide strands of each cable, at the center of each of the five spans.
3. Calculate the "center-of-gravity" position for each guide-strand group in each span.
4. Adjust the sag of each strand to bring it to the center-of gravity position in each span. This position was considered to represent the correct theoretical guide-strand sag in each span.

The maximum "spread" from the highest to the lowest strand at the span center, prior to adjustment, was found to be 1-3/4 in. (44 mm) in the main span, 3-1/2 in. (89 mm) in the side spans, and 3-3/4 in. (95 mm) in the back spans. Further, the maximum change of perpendicular sag needed to bring the guide strands to the center-of-gravity position in each span was found to be 15/16 in. (24 mm) for the main span, 2-1/16 in. (52 mm) for the side spans, and 2-1/16 in. (52 mm) for the back spans. These small adjustments testify to the accuracy of strand fabrication and to the validity of the setting-to-mark strand adjustment procedure, which was declared to be a success by all parties concerned. It seems doubtful that such accuracy in cable positioning could have been achieved using the standard sag-survey procedure.

With the first-layer strands in proper position in each cable, the strands in the second and subsequent layers were positioned to hang correctly in relation to the first layer, as is customary and proper for suspension bridge cable construction.

This example provides good illustration that the construction-engineering philosophy referred to as the shop-control procedure can be applied advantageously not only to typical rigid-type steel structures, such as continuous trusses and arches, but also to flexible-type structures, such as suspension bridges.

There is, however, an important caveat: the steelwork contractor must be a firm of suitable caliber and experience.

1.16   Example of Cable-Stayed Bridge Construction

In the case of cable-stayed bridges, the first of which were built in the 1950s, it appears that the governing construction engineering philosophy calls for field measurement and adjustment as the means for control of stay-cable and deck-structure geometry and stressing. For example, we have seen specifications calling for the completed bridge to meet the following geometric and stress requirements:

1. The deck elevation at midspan shall be within 12 in. (305 mm) of theoretical.
2. The deck profile at each cable attachment point shall be within 2 in. (50 mm) of a parabola passing through the actual (i.e., field-measured) midspan point.
3. Cable-stay tensions shall be within 5% of the "corrected theoretical" values.

July 6th, 1951
JJ:MM
C-1756

[To the design engineer]

Gentlemen: Attention of Mr. _____
                  Re: <u>Chesapeake Bay Bridge — Suspension Span Cables</u>

In our studies of the method of cable erection, we have arrived at the conclusion that setting of the guide strands to measured marks, instead of to surveyed sag, is a more satisfactory and more accurate method. Since such a procedure is not in accordance with the specifications, we wish to present for your consideration the reasoning which has led us to this conclusion, and to describe in outline form our proposed method of setting to marks.

On previous major suspension bridges, most of which have been built with parallel-wire instead of helical-strand cables, the thought has evidently been that setting the guide wire or guide strand to a computed sag, varying with the temperature, would be the most accurate method. This is associated with the fact that guide wires were never measured and marked to length. These established methods were carried over when strand-type cables came into use. An added reason may have been the knowledge that a small error in length results in a relatively large error in sag; and on the present structure the length-error to sag-error ratios are 1:2.4 and 1:1.5 for the main span and side spans, respectively.

However, the reading of the sag in the field is a very difficult operation because of the distances involved, the slopes of the side spans and backstays, the fact that even slight wind causes considerable motion to the guide strand, and for other practical reasons. We also believe that even though readings are made on cloudy days or at night, the actual temperature of all portions of the structure that will affect the sag cannot be accurately known. We are convinced that setting the guide strands according to the length marks thereon, which are placed under what amount to laboratory or ideal conditions at the manufacturing plant, will produce more accurate results than would field measurement of the sag.

To be specific, consider the case of field determination of sag in the main span, where it is necessary to establish accessible platforms, and an H.I. and a foresight somewhat below the desired sag elevation; and then to sight on the foresight and bring a target, hung from the guide strand, down to the line-of-sight. In the present case it is 1600 ft (488 m) to the foresight and 800 ft (244 m) to the target. Even if the line-of-sight were established just right, it would be only under perfect conditions of temperature and air — if indeed then — that such a survey would be precise. The difficulties are still greater in the side spans and back spans, where inclined lines-of-sight must be established by a series of offset measurements from distant bench marks. There is always the danger, particularly in the present location and at the time now scheduled, that days may be lost in waiting for the right conditions of weather to make an instrument survey feasible.

There is a second factor of doubt involved. The strand is measured under a known stress and at a known modulus, with "mechanical stretch" taken out. It is then reeled to a relatively small diameter and unreeled at the bridge site. Under its own weight, and until the full dead load has been applied, there is an indeterminable loss in mechanical set, or loss of modulus. A strand set to proper sag for the final modulus will accordingly be set too low, and the final cable will be below plan elevation. This possible error can only be on the side that is less desirable. Evidently, also, it could be on the order of 1-1/2 in. (40 mm) of sag increase for 1% of temporary reduction in modulus. If the strand were set to sag based on the assumed smaller modulus than will exist for the fully loaded condition, we doubt whether this smaller modulus could be chosen closely enough to ensure that the final sag would be correct. We are assured, however, by our manufacturing plant, that even though the modulus under bare-cable weight may be subject to unknown variation, the modulus which existed at the pre-stressing bed under the measuring tension will be duplicated when this same tension is reached under dead load. Therefore, if the guide stand is set to measured marks, the doubt as to modulus is eliminated.

FIGURE 1.7   Setting cable guide strands to marks.

A third source of error is temperature. In past practice the sag has been adjusted, by reference to a chart, in accordance with the existing temperature. Granted that the adjustment is made in the early morning (the fog having risen but the sun not), it is hard to conceive that the actual average temperature in 3955 ft (1205 m) of strand will be that recorded by any thermometer. The mainspan sag error is about 0.7 in. (18 mm) per deg C of temperature.

These conditions are all greatly improved at the strand pre-stressing bed. There seems to be no reason to doubt that the guide strands can be measured and marked to an insignificant degree of error, at a stipulated stress and under a well-soaked and determinable temperature. Any errors in sag level must result from something other than the measured length of the guide strand.

There is one indispensable condition which, however, holds for either method of setting. That is, that the total distance from anchorage to anchorage, and the total calculated length of strand under its own-weight stress, must agree within the limits of shimming provided in the anchorages. Therefore, this distance in the field must be checked to close agreement. While the measured length of strand will be calculated with precision, it is interesting to note that in this calculation, it is not essential that the modulus be known with exactness. The important factor is that the strand length under the final deadload stress will be calculated exactly; and since that length is measured under the corresponding average strand stress, knowledge of the modulus is not a consideration. If the modulus at deadload stress is not as assumed, the only effect will be a change of deflection under live load, and this is minor. We emphasize again that the stand length under dead load, and the length as measured in the prestressing bed, will be identical regardless of the modulus.

The calculations for the bare-cable position result in pulled-back positions for the tops of the towers and cable bents, in order to control the unbalanced forces tending to slip the strands in the saddles. These pullback distances may be slightly in error without the slipping forces overcoming friction and thereby becoming apparent. Such errors would affect the final sags of strands set to sag. However, they would have no effect on the final sags of strands set-to-mark at the saddles; these errors change the temporary strand sags only, and under final stress the sags and the shaft leans will be as called for by the design plans.

It sometimes has happened that a tower which at its base is square to the bridge axis, acquires a slight skew as it rises. The amount of this skew has never, so far as we know, been important. If it is disregarded and the guide strands are attached without any compensating change, then the final loading will, with virtual certainty, pull the tower square. All sources of possible maladjustment have now been discussed except one — the errors in the several span lengths at the base of the towers and bents. The intention is to recognize and accept these, by performing the appropriate check measurements; and to correct for them by slipping the guide strands designated amounts through the saddles such that the center-of-saddle mark on the strand will be offset by that same amount from the centerline of the saddle.

If we have left unexplained herein any factor that seems to you to render our procedure questionable, we are anxious to know of it and discuss it with you in the near future; and we will be glad to come to your offices for this purpose. The detailed preparations for observing strand sags would require considerable time, and we are not now doing any work along those lines.

Yours very truly,
Chief Engineer

FIGURE 1.7 (CONTINUED)   Setting cable guide strands to marks.

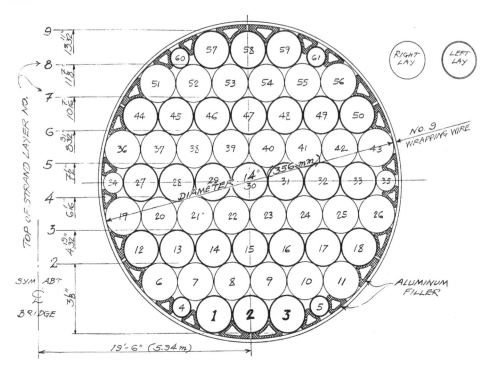

FIGURE 1.8 Main cable of first Chesapeake Bay suspension bridge, Maryland. Each cable consists of 61 helical-type bridge strands, 55 of 1-11/16 in. (43 mm) and 6 of 29/32 in. (23 mm) diameter. Strands 1, 2, and 3 were designated "guide strands" and were set to mark at each saddle and to normal shims at anchorages.

Such specification requirements introduce a number of problems of interpretation, field measurement, calculation, and field correction procedure, such as the following:

1. Interpretation:
   - The specifications are silent with respect to transverse elevation differentials. Therefore, two deck-profile control parabolas are presumably needed, one for each side of the bridge.

2. Field measurement of actual deck profile:
   - The temperature will be neither constant nor uniform throughout the structure during the survey work.
   - The survey procedure itself will introduce some inherent error..

3. Field measurement of cable-stay tensions:
   - Hydraulic jacks, if used, are not likely to be accurate within 2%, perhaps even 5%; further, the exact point of "lift off" will be uncertain.
   - Other procedures for measuring cable tension, such as vibration or strain gaging, do not appear to define tensions within about 5%.
   - All cable tensions cannot be measured simultaneously; an extended period will be needed, during which conditions will vary and introduce additional errors.

4. Calculation of "actual" bridge profile and cable tensions:
   - Field-measured data must be transformed by calculation into "corrected actual" bridge profiles and cable tensions, at normal temperature and without erection loads.
   - Actual dead weights of structural components can differ by perhaps 2% from nominal weights, while temporary erection loads probably cannot be known within about 5%.
   - The actual temperature of structural components will be uncertain and not uniform.

FIGURE 1.9 Cable-stayed orthotropic-steel-deck bridge over Mississippi River at Luling, La., 1982; view looking northeast. The main span is 1222 ft (372 m); the A-frame towers are 350 ft (107 m) high. A barge-mounted ringer derrick erected the main steelwork, using a 340 ft (104 m) boom with a 120 ft (37 m) jib to erect tower components weighing up to 183 tons, and using a shorter boom for deck components. Cable stays at the ends of projecting cross girders are permanent; others are temporary erection stays. Girder section 16-west of north portion of bridge, erected a few days previously, is projecting at left; companion girder section 16-east is on barge ready for erection (see Figure 1.10).

- The mathematical model itself will introduce additional error.

5. "Target condition" of bridge:

  - The "target condition" to be achieved by field adjustment will differ from the geometric condition, because of the absence of the deck wearing surface and other such components; it must therefore be calculated, introducing additional error.

6. Determining field corrections to be carried out by erector, to transform "corrected actual" bridge into "target condition" bridge:

  - The bridge structure is highly redundant, and changing any one cable tension will send geometric and cable-tension changes throughout the structure. Thus, an iterative correction procedure will be needed.

It seems likely that the total effect of all these practical factors could easily be sufficient to render ineffective the contractor's attempts to fine tune the geometry and stressing of the as-erected structure in order to bring it into agreement with the calculated bridge target condition. Further, there can be no assurance that the specifications requirements for the deck-profile geometry and cable-stay tensions are even compatible; it seems likely that *either* the deck geometry *or* the cable tensions may be achieved, but not *both*.

Specifications clauses of the type cited seem clearly to constitute unwarranted and unnecessary field-adjustment requirements. Such clauses are typically set forth by bridge designers who have great confidence in computer-generated calculation, but do not have a sufficient background in and understanding of the practical factors associated with steel bridge construction. Experience has shown that field proce-

FIGURE 1.10 Luling Bridge deck steelwork erection, 1982; view looking northeast (refer to Figure 1.9). The twin box girders are 14 ft (4.3 m) deep; the deck plate is 7/16 in. (11 mm) thick. Girder section 16-east is being raised into position (lower right) and will be secured by large-pin hinge bars prior to fairing-up of joint holes and permanent bolting. Temporary erection stays are jumped forward as girder erection progresses.

dures for major bridges developed unilaterally by design engineers should be reviewed carefully to determine whether they are practical and desirable and will in fact achieve the desired objectives.

In view of all these considerations, the question comes forward as to what design and construction principles should be followed to ensure that the deadload geometry and stressing of steel cable-stayed bridges will fall within acceptable limits. Consistent with the general construction-engineering procedures recommended for other types of bridges, we should abandon reliance on field measurements followed by adjustments of geometry and stressing, and instead place prime reliance on proper geometric control of bridge components during fabrication, followed by accurate erection evolutions as the work goes forward in the field.

Accordingly, the proper construction procedure for cable-stayed steel bridges can be summarized as follows:

1. Determine the actual bridge baseline lengths and pier-top elevations to a high degree of accuracy.
2. Fabricate the bridge towers, cables, and girders to a high degree of geometric precision.
3. Determine, in the fabricating shop, the final residual errors in critical fabricated dimensions, including cable-stay lengths after socketing, and positions of socket bearing surfaces or pinholes.
4. Determine "corrected theoretical" positioning for each individual cable stay.
5. During erection, bring all tower and girder structural joints into shop-fabricated alignment, with fair holes, etc.
6. At the appropriate erection stages, install "corrected theoretical" positional for each cable stay.
7. With the structure in the all-steel-erected condition (or other appropriate designated condition), check it over carefully to determine whether any significant geometric or other discrepancies are in evidence. If there are none, declare conditions acceptable and continue with erection.

This construction engineering philosophy can be summarized by stating that if the steelwork fabrication and erection are properly engineered and carried out, the geometry and stressing of the completed structure will fall within acceptable limits; whereas, if the fabrication and erection are not properly done, corrective measurements and adjustments attempted in the field are not likely to improve the structure, or even to prove satisfactory. Accordingly, in constructing steel cable-stayed bridges we should place full reliance on accurate shop fabrication and on controlled field erection, just as is done on other types of steel bridges, rather than attempting to make measurements and adjustments in the field to compensate for inadequate fabrication and erection.

1.17 Field Checking at Critical Erection Stages

As has been stated previously, the best governing procedure for steel bridge construction is generally the shop control procedure, wherein full reliance is placed on accurate fabrication of the bridge components as the basis for the integrity of the completed structure. However, this philosophy does not rule out the desirability of certain checks in the field as erection goes forward, with the objective of providing assurance that the work is on target and no significant errors have been introduced.

It would be impossible to catalog those cases during steel bridge construction where a field check might be desirable; such cases will generally suggest themselves as the construction engineering studies progress. We will only comment that these field-check cases, and the procedures to be used, should be looked at carefully, and even skeptically, to make certain that the measurements will be both desirable and practical, producing meaningful information that can be used to augment job integrity.

1.18 Determination of Erection Strength Adequacy

Quite commonly, bridge member forces during the erection stages will be altogether different from those that will prevail in the completed structure. At each critical erection stage the bridge members must be reviewed for strength and stability, to ensure structural integrity as the work goes forward. Such a construction engineering review is typically the responsibility of the steelwork erector, who carries out thorough erection studies of the structure and calls for strengthening or stabilizing of members as needed. The erector submits the studies and recommendations to the design engineer for review and comment, but normally the full responsibility for steelwork structural integrity during erection rests with the erector.

In the U.S., bridgework design specifications commonly require that stresses in steel structures under erection shall not exceed certain multiples of design allowable stresses. Although this type of erection stress limitation is probably safe for most steel structures under ordinary conditions, it is not necessarily adequate for the control of the erection stressing of large monumental-type bridges. The key point to be understood here is that fundamentally there is no logical fixed relationship between design allowable stresses, which are based upon somewhat uncertain long-term service loading requirements along with some degree of assumed structural deterioration, and stresses that are safe and economical during the bridge erection stages, where loads and their locations are normally well defined and the structural material is in new condition. Clearly, the basic premises of the two situations are significantly different, and "factored design stresses" must therefore be considered unreliable as a basis for evaluating erection safety.

There is yet a further problem with factored design stresses. Large truss-type bridges in various erection stages may undergo deflections and distortions that are substantial compared with those occurring under service conditions, thereby introducing apprehension regarding the effect of the secondary bending stresses that result from joint rigidity.

Recognizing these basic considerations, the engineering department of a major U.S. steelwork contractor went forward in the early 1970s to develop a logical philosophy for erection strength appraisal of large structural steel frameworks, with particular reference to long-span bridges, and implemented this philosophy with a stress analysis procedure. The effort was successful and the results were reported in a paper published by the American Society of Civil Engineers in 1977 [6]. This stress analysis procedure, designated the erection rating factor (ERF) procedure, is founded directly upon basic structural principles, rather than on bridge-member design specifications, which are essentially irrelevant to the problem of erection stressing.

FIGURE 1.11 First Quebec railway cantilever bridge, 23 August 1907. Cantilever erection of south main span is shown six days before collapse. The tower traveler erected the anchor span (on falsework) and then the cantilever arm; then erected the top-chord traveler, which is shown erecting suspended span at end of cantilever arm. The main span of 1800 ft (549 m) was the world's longest of any type. The sidespan bottom chords second from pier (arrow) failed in compression because latticing connecting chord corner angles was deficient under secondary bending conditions.

It may be noted that a significant inducement toward development of the ERF procedure was the failure of the first Quebec cantilever bridge in 1907 (see Figures 1.11 and 1.12). It was quite obvious that evaluation of the structural safety of the Quebec bridge at advanced cantilever erection stages such as that portrayed in Figure 1.11, by means of the factored-design-stress procedure, would inspire no confidence and would not be justifiable.

The erection rating factor (ERF) procedure for a truss bridge can be summarized as follows:

1. Assume either (a) pin-ended members (no secondary bending), (b) plane-frame action (rigid truss joints, secondary bending in one plane), or (c) space-frame action (bracing-member joints also rigid, secondary bending in two planes), as engineering judgment dictates.
2. Determine, for each designated erection stage, the member primary forces (axial) and secondary forces (bending) attributable to gravity loads and wind loads.
3. Compute the member stresses induced by the combined erection axial forces and bending moments.
4. Compute the ERF for each member at three or five locations: at the middle of the member; at each joint, inside the gusset plates (usually at the first row of bolts); and, where upset member plates or gusset plates are used, at the stepped-down cross-section outside each joint.
5. Determine the minimum computed ERF for each member and compare it with the stipulated minimum value.
6. Where the computed minimum ERF equals or exceeds the stipulated minimum value, the member is considered satisfactory. Where it is less, the member may be inadequate; reevaluate the critical part of it in greater detail and recalculate the ERF for further comparison with the stipulated minimum. (Initially calculated values can often be increased significantly.)
7. Where the computed minimum ERF remains less than the stipulated minimum value, strengthen the member as required.

FIGURE 1.12 Wreckage of south anchor span of first Quebec railway cantilever bridge, 1907. View looking north from south shore a few days after the collapse on 29 August 1907, the worst disaster in the history of bridge construction. About 20,000 tons of steelwork fell into the St. Lawrence River, and 75 workmen lost their lives.

Note that member forces attributable to wind are treated the same as those attributable to gravity loads. The old concept of "increased allowable stresses" for wind is not considered to be valid for erection conditions and is not used in the ERF procedure. Maximum acceptable ℓ/r and b/t values are included in the criteria. ERFs for members subjected to secondary bending moments are calculated using interaction equations.

1.19 Philosophy of the Erection Rating Factor

In order that the structural integrity and reliability of a steel framework can be maintained throughout the erection program, the minimum probable (or "minimum characteristic") strength value of each member must necessarily be no less than the maximum probable (or "maximum characteristic") force value, under the most adverse erection condition. In other words, the following relationship is required:

$$S - \Delta S \geq F + \Delta F \qquad (1.1)$$

where

S = computed or nominal strength value for the member
ΔS = maximum probable member strength underrun from the computed or nominal value
F = computed or nominal force value for the member
ΔF = maximum probable member force overrun from the computed or nominal value

Equation 1.1 states that in the event the actual strength of the structural member is less than the nominal strength, S, by an amount ΔS, while at same time the actual force in the member is greater than the nominal force, F, by an amount ΔF, the member strength will still be no less than the member force, and so the member will not fail during erection. This equation provides a direct appraisal of erection realities, in contrast to the allowable-stress approach based on factored design stresses.

Proceeding now to rearrange the terms in Equation 1.1, we find that

$$S\left(1-\frac{\Delta S}{S}\right) \geq F\left(1+\frac{\Delta F}{F}\right); \quad \frac{S}{F} \geq \frac{1+\frac{\Delta F}{F}}{1-\frac{\Delta S}{S}} \tag{1.2}$$

The ERF is now defined as

$$ERF \equiv \frac{S}{F} \tag{1.3}$$

that is, the nominal strength value, S, of the member divided by its nominal force value, F. Thus, for erection structural integrity and reliability to be maintained, it is necessary that

$$ERF \geq \frac{1+\frac{\Delta F}{F}}{1-\frac{\Delta S}{S}} \tag{1.4}$$

1.20 Minimum Erection Rating Factors

In view of possible errors in (1) the assumed weight of permanent structural components, (2) the assumed weight and positioning of erection equipment, and (3) the mathematical models assumed for purposes of erection structural analysis, it is reasonable to assume that the actual member force for a given erection condition may exceed the computed force value by as much as 10%; that is, it is reasonable to take $\Delta F/F$ as equal to 0.10.

For tension members, uncertainties in (1) the area of the cross-section, (2) the strength of the material, and (3) the member workmanship, indicate that the actual member strength may be up to 15% less than the computed value; that is, $\Delta S/S$ can reasonably be taken as equal to 0.15. The additional uncertainties associated with compression member strength suggest that $\Delta S/S$ be taken as 0.25 for those members. Placing these values into Equation 1.4, we obtain the following minimum ERFs:

Tension members: $ERF_{t,min} = (1+0.10)/(1-0.15)$

$= 1.294$, say 1.30

Compression members: $ERF_{c,min} = (1+0.10)/(1-0.25)$

$= 1.467$, say 1.45

The proper interpretation of these expressions is that if, for a given tension (compression) member, the ERF is calculated as 1.30 (1.45) or more, the member can be declared safe for the particular erection condition. Note that higher, or lower, values of erection rating factors may be selected if conditions warrant.

The minimum ERFs determined as indicated are based on experience and judgment, guided by analysis and test results. They do not reflect any specific probabilities of failure and thus are not based on the concept of an acceptable risk of failure, which might be considered the key to a totally rational approach to structural safety. This possible shortcoming in the ERF procedure might be at least partially overcome by evaluating the parameters $\Delta F/F$ and $\Delta S/S$ on a statistical basis; however, this would involve a considerable effort, and it might not even produce significant results.

It is important to recognize that the ERF procedure for determining erection strength adequacy is based directly on fundamental strength and stability criteria, rather than being only indirectly related to such criteria through the medium of a design specification. Thus, the procedure gives uniform results for the erection rating of framed structural members irrespective of the specification that was used to design the members. Obviously, the end use of the completed structure is irrelevant to its strength adequacy during the erection configurations, and therefore the design specification should not be brought into the picture as the basis for erection appraisal.

Experience with application of the ERF procedure to long-span truss bridges has shown that it places the erection engineer in much better contact with the physical significance of the analysis than can be obtained by using the factored-design-stress procedure. Further, the ERF procedure takes account of secondary stresses, which have generally been neglected in erection stress analysis.

Although the ERF procedure was prepared for application to truss bridge members, the simple governing structural principle set forth by Equation 1.1 could readily be applied to bridge members and components of any type.

1.21 Deficiencies of Typical Construction Procedure Drawings and Instructions

At this stage of the review it is appropriate to bring forward a key problem in the realm of bridge construction engineering: the strong tendency for construction procedure drawings to be insufficiently clear, and for step-by-step instructions to be either lacking or less than definitive. As a result of these deficiencies it is not uncommon to find the contractor's shop and field evolutions to be going along under something less than suitable control.

Shop and field operations personnel who are in a position to speak frankly to construction engineers will sometimes let them know that procedure drawings and instructions often need to be clarified and upgraded. This is a pervasive problem, and it results from two prime causes: (1) the fabrication and erection

FIGURE 1.13 Visiting the work site. It is of first-order importance for bridge construction engineers to visit the site regularly and confer with the job superintendent and his foremen regarding practical considerations. Construction engineers have much to learn from the work forces in shop and field, and vice versa. (Courtesy of Bethlehem Steel Corporation.)

engineers responsible for drawings and instructions do not have adequate on-the-job experience, and (2) they are not sufficiently skilled in the art of setting forth on the documents, clearly and concisely, exactly what is to be done by the operations forces — and, sometimes of equal importance, what *is not* to be done.

This matter of clear and concise construction procedure drawings and instructions may appear to be a pedestrian matter, but it is decidedly not. *It is a key issue of utmost importance to the success of steel bridge construction.*

1.22   Shop and Field Liaison by Construction Engineers

In addition to the need for well-prepared construction procedure drawings and instructions, it is essential for the staff engineers carrying out construction engineering to set up good working relations with the shop and field production forces, and to visit the work sites and establish effective communication with the personnel responsible for accomplishing what is shown on the documents.

Construction engineers should review each projected operation in detail with the work forces, and upgrade the procedure drawings and instructions as necessary, as the work goes forward. Further, engineers should be present at the work sites during critical stages of fabrication and erection. As a component of these site visits, the engineers should organize special meetings of key production personnel to go over critical operations in detail — complete with slides and blackboard as needed — thereby providing the work forces with opportunities to ask questions and discuss procedures and potential problems, and providing engineers the opportunity to determine how well the work forces understand the operations to be carried out.

This matter of liaison between the office and the work sites — like the preceding issue of clear construction procedure documents — may appear to be somewhat prosaic; again, however, *it is a matter of paramount importance.* Failure to attend to these two key issues constitutes a serious problem in steel bridge construction, and opens the door to high costs and delays, and even to erection accidents.

1.23   Comprehensive Bridge Erection-Engineering Specifications

The erection rating factor (ERF) procedure for determination of erection strength adequacy, as set forth heretofore for bridge trusswork, could readily be extended to cover bridge members and components of any type under erection loading conditions. Bridge construction engineers should work toward this objective, in order to release erection strength appraisal from the limitations of the commonly used factored-design-stress procedure.

Looking still further ahead, it is apparent that there is need in the bridge engineering profession for comprehensive erection engineering specifications for steel bridge construction. Such specifications should include guidelines for such matters as devising and evaluating erection schemes, determining erection loads, evaluating erection strength adequacy of all types of bridge members and components, designing erection equipment, and designing temporary erection members such as falsework, tiedowns, tiebacks, and jacking struts. The specifications might also cover contractual considerations associated with construction engineering.

The key point to be recognized here is that the use of bridge *design* specifications as the basis for erection engineering studies, as is currently the custom, is not appropriate. Erection engineering is a related but different discipline, and should have its own specifications. However, given the current fragmented state of construction engineering in the U.S. (refer to Section 1.26), it is difficult to envision how such erection engineering specifications could be prepared. Proprietary considerations associated with each erection firm's experience and procedures could constitute an additional obstacle.

1.24   Standard Conditions for Contracting

A further basic problem in respect to the future of steel bridge construction in the U.S. lies in the absence of standard conditions for contracting.

On through the 19th century both the design and the construction of a major bridge in the U.S. were frequently the responsibility of a single prominent engineer, who could readily direct and coordinate the work and resolve problems equitably. Then, over, the first thirty years or so of the 20th century this system was progressively displaced by the practice of competitive bidding on plans and specifications prepared by a design engineer retained by the owner. As a result the responsibility for the structure previously carried by the designer-builder became divided, with the designer taking responsibility for service integrity of the completed structure while prime responsibility for structural adequacy and safety during construction was assumed by the contractor. Full control over the preparation of the plans and specifications — the contract documents — was retained by the design engineer.

This divided responsibility has resulted in contract documents that may not be altogether equitable, since the designer is inevitably under pressure to look after the immediate financial interests of his client, the owner. Documents prepared by only one party to a contract can hardly be expected to reflect the appropriate interest of the other party. However, until about mid-20th-century design and construction responsibilities for major bridgework, although divided between the design engineer and the construction engineer, were nonetheless usually under the control of leading members of the bridge engineering profession who were able to command the level of communication and cooperation needed for resolution of inevitable differences of opinion within a framework of equity and good will.

Since the 1970s there has been a trend away from this traditional system of control. The business and management aspects of design firms have become increasingly important, while at the same time steel-work construction firms have become more oriented toward commercial and legal considerations. Professional design and construction engineers have lost stature correspondingly. As a result of these adverse trends, bridgework specifications are being ever more stringently drawn, bidding practices are becoming increasingly aggressive, claims for extra reimbursement are proliferating, insurance costs for all concerned are rising, and control of bridge engineering and construction is being influenced to an increasing extent by administrators and attorneys. These developments have not benefited the bridge owners, the design engineering profession, the steelwork construction industry, or the public — which must ultimately pay all the costs of bridge construction.

It seems clear that in order to move forward out of this unsatisfactory state of affairs, a comprehensive set of standard conditions for contracting should be developed to serve as a core document for civil engineering construction — a document that would require only the addition of special provisions in order to constitute the basic specifications for any major bridge construction project. Such standard conditions would have to be prepared "off line" by a group of high-level engineering delegates having well established engineering credentials.

A core contract document such as the one proposed has been in general use in Great Britain since 1945, when the first edition of the *Conditions of Contract and Forms of Tender, Agreement and Bond for Use in Connection with Works of Civil Engineering Construction* was published by The Institution of Civil Engineers (ICE). This document, known informally as *The ICE Conditions of Contract*, is now in its 6th edition [1]. It is kept under review and revised as necessary by a permanent Joint Contracts Committee consisting of delegates from The Institution of Civil Engineers, The Association of Consulting Engineers, and The Federation of Civil Engineering Contractors. This document is used as the basis for the majority of works of civil engineering construction that are contracted in Great Britain, including steel bridges.

Further comments on the perceived need for U.S. standard conditions for contracting can be found in [7].

1.25 Design-and-Construct

As has been mentioned, design-and-construct was common practice in the U.S. during the 19th century. Probably the most notable example was the Brooklyn Bridge, where the designer-builders were John A. Roebling and his son Washington A. Roebling. Construction of the Brooklyn Bridge was begun in 1869 and completed in 1883. Design-and-construct continued in use through the early years of the 20th century; the most prominent example from that era may be the Ambassador suspension bridge between

Detroit, Michigan, and Windsor, Ontario, Canada, completed in 1929. The Ambassador Bridge was designed and built by the McClintic-Marshall Construction Co., Jonathan Jones, chief engineer; it has an 1850 ft main span, at that time the world's record single span.

Design-and-construct has not been used for a major steel bridge in the U.S. since the Ambassador Bridge. However, the procedure has seen significant use throughout the 20th century for bridges in other countries and particularly in Europe; and most recently design-construct-operate-maintain has come into the picture. Whether these procedures will find significant application in the U.S. remains to be seen.

The advantages of design-and-construct are readily apparent:

1. More prospective designs are likely to come forward, than when designs are obtained from only a single organization.
2. Competitive designs are submitted at a preliminary level, making it possible for the owner to provide some input to the selected design between the preliminary stage and design completion.
3. The owner knows the price of the project at the time the preliminary design is selected, as compared with design-bid where the price is not known until the design is completed and bids are received.
4. As the project goes forward the owner deals with only a single entity, thereby reducing and simplifying his administrative effort.
5. The design-and-construct team members must work effectively together, eliminating the antagonisms and confrontations that can occur on a design-bid project.

A key requirement in the design-and-construct system for a project is the meticulous preparation of the request-for-proposals (RFP), which should cover the following essentials in suitable detail and clarity:

1. Description of project to be constructed.
2. Scope of work.
3. Structural component types and characteristics: which are required, which are acceptable, and which are not acceptable.
4. Minimum percentages of design and construction work that must be performed by the team's own forces.
5. Work schedules; time incentives and disincentives.
6. Procedure to be followed when actual conditions are found to differ from those assumed.
7. Quality control and quality assurance factors.
8. Owner's approval prerogatives during final-design stage and construction stage.
9. Applicable local, state, and federal regulations.
10. Performance and payment bonding requirements.
11. Warranty requirements.
12. Owner's procedure for final approval of completed project.

In preparing the RFP, the owner should muster all necessary resources from both inside and outside his organization. Political considerations should be given due attention. Document drafts should receive the appropriate reviews, and an RFP brought forward that is in near-final condition. Then, at the start of the contracting process, the owner will typically proceed as follows:

1. Announce the project and invite prospective teams to submit qualifications.
2. Prequalify a small number of teams, perhaps three to five, and send the draft RFP to each.
3. Hold a meeting with the prequalified teams for informal exchange of information and to discuss questions.
4. Prepare the final RFP and issue it to each prequalified team, and announce the date on which proposals will be due.

The owner will customarily call for the proposals to be submitted in two separate components: the design component, showing the preliminary design carried to about the 25% level; and the monetary component, stating the lump-sum bid. Before the bids are opened the owner will typically carry out a scoring process for the preliminary designs, not identifying the teams with their designs, using a 10 point or 100 point grading scale and giving consideration to the following factors:

1. Quality of the design.
2. Bridge aesthetics.
3. Fabrication and erection feasibility and reliability.
4. Construction safety aspects.
5. Warranty and long-term maintenance considerations.
6. User costs.

Using these and other such scoring factors (which can be assigned weights if desired), a final overall design score is assigned to each preliminary design. Then the lump-sum bids are opened. A typical procedure is to divide each team's bid price by its design score, yielding and overall price rating, and to award the contract to the design-and-construct team having the lowest price rating.

Following the contract award the successful team will proceed to bring its preliminary design up to the final-design level, with no site work permitted during this interval. It is customary for the owner to award each unsuccessful submitting team a stipend to partially offset the costs of proposal preparation.

1.26 Construction Engineering Procedures and Practices — The Future

The many existing differences of opinion and procedures in respect to proper governance of steelwork fabrication and erection for major steel bridges raises the question: How do proper bridge construction guidelines come into existence and find their way into practice and into bridge specifications? Looking back over the period roughly from 1900 to 1975, we find that the major steelwork construction companies in the U.S. developed and maintained competent engineering departments that planned and engineered large bridges (and smaller ones as well) through the fabrication and erection processes with a high degree of proficiency. Traditionally, the steelwork contractor's engineers worked in cooperation with design-office engineers to develop the full range of bridgework technical factors, including construction procedure and practices.

However, times have changed; since the 1970s major steel bridge contractors have all but disappeared in the U.S., and further, very few bridge design offices have on their staffs engineers experienced in fabrication and erection engineering. As a result, construction engineering often receives less attention and effort than it needs and deserves, and this is not a good omen for the future of the design and construction of large bridges in the U.S.

Bridge construction engineering is not a subject that is or can be taught in the classroom; it must be learned on the job with major steelwork contractors. The best route for an aspiring young construction engineer is to spend significant amounts of time in the fabricating shop and at the bridge site, interspersed with time doing construction-engineering technical work in the office. It has been pointed out previously that although construction engineering and design engineering are related, they constitute different practices and require diverse backgrounds and experience. Design engineering can essentially be learned in the design office; construction engineering, however, cannot — it requires a background of experience at work sites. Such experience, it may be noted, is valuable also for design engineers; however, it is not as necessary for them as it is for construction engineers.

The training of future steelwork construction engineers in the U.S. will be handicapped by the demise of the "Big Two" steelwork contractors in the 1970s. Regrettably, it appears that surviving steelwork contractors in the U.S. generally do not have the resources for supporting strong engineering departments, and so there is some question as to where the next generation of steel bridge construction engineers in the U.S. will be coming from.

1.27 Concluding Comments

In closing this review of steel bridge construction it is appropriate to quote from the work of an illustrious British engineer, teacher, and author, the late Sir Alfred Pugsley [15]:

> A further crop of [bridge] accidents arose last century from overloading by traffic of various kinds, but as we have seen, engineers today concentrate much of their effort to ensure that a margin of strength is provided against this eventuality. But there is one type of collapse that occurs almost as frequently today as it has over the centuries: collapse at a late stage of erection.

> The erection of a bridge has always presented its special perils and, in spite of ever-increasing care over the centuries, few great bridges have been built without loss of life. Quite apart from the vagaries of human error, with nearly all bridges there comes a critical time near completion when the success of the bridge hinges on some special operation. Among such are ... the fitting of a last section ... in a steel arch, the insertion of the closing central [members] in a cantilever bridge, and the lifting of the roadway deck [structure] into position on a suspension bridge. And there have been major accidents in many such cases. It may be wondered why, if such critical circumstances are well known to arise, adequate care is not taken to prevent an accident. Special care is of course taken, but there are often reasons why there may still be "a slip bewixt cup and lip". Such operations commonly involve unusually close cooperation between constructors and designers, and between every grade of staff, from the laborers to the designers and directors concerned; and this may put a strain on the design skill, on detailed inspection, and on practical leadership that is enough to exhaust even a Brunel.

> In such circumstances it does well to ... recall [the] dictum ... that "it is essential not to have faith in human nature. Such faith is a recent heresy and a very disastrous one." One must rely heavily on the lessons of past experience in the profession. Some of this experience is embodied in professional papers describing erection processes, often (and particularly to young engineers) superficially uninteresting. Some is crystallized in organizational habits, such as the appointment of resident engineers from both the contracting and [design] sides. And some in precautions I have myself endeavored to list. ...

> It is an easy matter to list such precautions and warnings, but quite another for the senior engineers responsible for the completion of a bridge to stand their ground in real life. This is an area of our subject that depends in a very real sense on the personal qualities of bridge engineers. ... At bottom, the safety of our bridges depends heavily upon the integrity of our engineers, particularly the leading ones.

1.28   Further Illustrations of Bridges under Construction, Showing Erection Methods

FIGURE 1.14   Royal Albert Bridge across River Tamar, Saltash, England, 1857. The two 455 ft (139 m) main spans, each weighing 1060 tons, were constructed on shore, floated out on pairs of barges, and hoisted about 100 ft (30 m) to their final position using hydraulic jacks. Pier masonry was built up after each 3 ft (1 m) lift.

FIGURE 1.15 Eads Bridge across the Mississippi River, St. Louis, Mo., 1873. The first important metal arch bridge in the U.S., it is supported by four planes of hingeless trussed arches having chrome-steel tubular chords. Spans are 502-520-502 ft (153-158-153 m). During erection, arch trusses were tied back by cables passing over temporary towers built on the piers. Arch ribs were packed in ice to effect closure.

GLASGOW STEEL BRIDGE,
CHICAGO AND ALTON RAILROAD.
April 8, 1879.
—
WM. SOOY SMITH,
Engineer

FIGURE 1.16 Glasgow (Missouri) railway truss bridge, 1879. Erection on full supporting falsework was common-place in the 19th century. The world's first all-steel bridge, with five 315 ft (96 m) through-truss simple spans, crossed the Missouri River.

FIGURE 1.17 Niagara River railway cantilever truss bridge, near Niagara Falls, New York, 1883. Massive wood erection traveler constructed side span on falsework, then cantilevered half of main span to midspan. Erection of other half of bridge was similar. First modern-type cantilever bridge, with 470 ft (143 m) clear main span having a 120 ft (37 m) center suspended span.

᷒. The massive cantilevers of the Forth bridge, shown under erection, were conceived in the shadow of the Tay bridge disaster.

FIGURE 1.18 Construction of monumental Forth Bridge, Scotland, 1888. Numerous small movable booms were used, along with erection travelers for cantilevering the two 1710 ft (521 m) main spans. The main compression members are tubes 12 ft (3.65 m) in diameter; many other members are also tubular. Total steelwork weight is 51,000 tons. Records are not clear regarding such essentials as cambering and field fitting of individual members in this heavily redundant railway bridge. The Forth is arguably the world's greatest steel structure.

FIGURE 1.19 Pecos River railway viaduct, Texas, 1892. Erection by massive steam-powered wood traveler having many sets of falls and very long reach. Cantilever-truss main span has 185 ft (56 m) clear opening.

FIGURE 1.20 Raising of suspended span, Carquinez Strait Bridge, California, 1927. The 433 ft (132 m) suspended span, weighting 650 tons, was raised into position in 35 min., driven by four counterweight boxes having a total weight of 740 tons.

FIGURE 1.21 First Cooper River cantilever bridge, Charleston, S.C., 1929. Erection travelers constructed 450 ft (137 m) side spans on falsework, then went on to erect 1050 ft (320 m) main span (including 437.5 ft [133 m] suspended span) by cantilevering to midspan.

FIGURE 1.22 Erecting south tower of Golden Gate Bridge, San Francisco, 1935. A creeper traveler with two 90 ft (27 m) booms erects a tier of tower cells for each leg, then is jumped to the top of that tier and proceeds to erect the next tier. The tower legs are 90 ft (27 m) center-to-center and 690 ft (210 m) high. When the traveler completed the north tower (in background) it erected a Chicago boom on the west tower leg, which dismantled the creeper, erected tower-top bracing, and erected two small derricks (one shown) to service cable erection. Each tower contains 22,200 tons of steelwork.

FIGURE 1.23 Balanced-cantilever erection, Governor O.K. Allen railway/highway cantilever bridge, Baton Rouge, La., 1939. First use of long balanced-cantilever erection procedure in the U.S. On each pier 650 ft (198 m) of steelwork, about 4000 tons, was balanced on the 40 ft (12 m) base formed by a sloping falsework bent. The compression load at the top of the falsework bent was measured at frequent intervals and adjusted by positioning a counterweight car running at bottom-chord level. The main spans are 848-650-848 ft (258-198-258 m); 650 ft span shown. (Courtesy of Bethlehem Steel Corporation.)

FIGURE 1.24 Tower erection, second Tacoma Narrows Bridge, Washington, 1949. This bridge replaced first Tacoma Narrows bridge, which blew down in a 40 mph (18 m/sec) wind in 1940. Tower legs are 60 ft (18 m) on centers and 462 ft (141 m) high. Creeper traveler is shown erecting west tower, in background. On east tower, creeper erected a Chicago boom at top of south leg; this boom dismantled creeper, then erected tower-top bracing and a stiffleg derrick, which proceeded to dismantle Chicago boom. Tower manhoist can be seen at second-from-topmost landing platform. Riveting cages are approaching top of tower. Note tower-base erection kneebraces, required to ensure tower stability in free-standing condition (see Figure 1.27).

FIGURE 1.25 Aerial spinning of parallel-wire main cables, second Tacoma Narrows suspension bridge, Washington, 1949. Each main cable consists of 8702 parallel galvanized high-strength wires of 0.196 in (4.98 mm) diameter, laid up as 19 strands of mostly 460 wires each. Following compaction the cable became a solid round mass of wires with a diameter of 20-1/4 in. (514 mm).

FIGURE 1.25A Tramway starts across from east anchorage carrying two wire loops. Three 460-wire strands have been spun, with two more under construction. Tramway spinning wheels pull wire loops across the three spans from east anchorage to west anchorage. Suspended footbridges provide access to cables. Spinning goes on 24 hours per day.

FIGURE 1.25B Tramway arrives at west anchorage. Wire loops shown in Figure 1.25A are removed from spinning wheels and placed around strand shoes at west anchorage. This tramway then returns empty to east anchorage, while tramway for other "leg" of endless hauling rope brings two wire loops across for second strand that is under construction for this cable.

FIGURE 1.26 Cable-spinning procedure for constructing suspension bridge parallel-wire main cables, showing details of aerial spinning method for forming individual 5 mm wires into strands containing 400 to 500 wires. Each wire loop is erected as shown in Figure 1.26A (refer to Figure 1.25), then adjusted to the correct sag as shown in Figure 1.26B. Each completed strand is banded with tape, then adjusted to the correct sag in each span. With all strands in place, they are compacted to form a solid round homogeneous mass of cable wires. The aerial spinning method was developed by John Roebling in the mid-19th century.

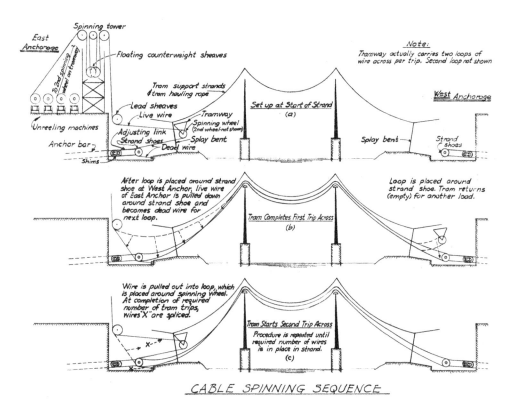

FIGURE 1.26A Erection of individual wire loops.

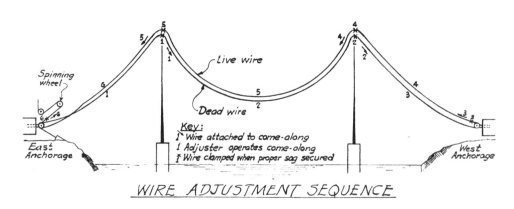

FIGURE 1.26B Adjustment of individual wire loops.

FIGURE 1.27 Erection of suspended deck steelwork, second Tacoma Narrows Bridge, Washington, 1950. Chicago boom on tower raises deck steelwork components to deck level, where they are transported to deck travelers by material cars. Each truss double panel is connected at top-chord level to previously erected trusses, and left open at bottom-chord level to permit temporary upward deck curvature, which results from partial loading condition of main suspension cables. Main span (at right) is 2800 ft (853 m), and side spans are 1100 ft (335 m). Stiffening trusses are 33 ft (10 m) deep and 60 ft (18 m) on centers. Tower-base kneebraces (see Figure 1.24) show clearly here.

(a)   (b)   (c)

FIGURE 1.28 Moving deck traveler forward, second Tacoma Narrows Bridge, Washington, 1950. Traveler pulling falls leadline passes around sheave beams at forward end of stringers, and is attached to front of material car (at left). Material car is pulled back toward tower, advancing traveler two panels to its new position at end of deck steelwork. Arrows show successive positions of material car. (a) Traveler at start of move, (b) traveler advanced one panel, and (c) traveler at end of move.

FIGURE 1.29 Erecting closing girder sections of Passaic River Bridge, New Jersey Turnpike, 1951. Huge double-boom travelers, each weighing 270 tons, erect closing plate girders of the 375 ft (114 m) main span. Closing girders are 14 ft (4.3 m) deep and 115 ft (35 m) long and weigh 146 tons each. Sidewise entry was required (as shown) because of long projecting splice material. Longitudinal motion was provided at one pier, where girders were jacked to effect closure. Closing girders were laterally stable without floor steel fill-in, such that derrick falls could be released immediately. (Courtesy of Bethlehem Steel Corporation.)

(a)   (b)

FIGURE 1.30 Floating-in erection of a truss span, first Chesapeake Bay Bridge, Maryland, 1951. Erected 300 ft (91 m) deck-truss spans form erection dock, providing a work platform for two derrick travelers. A permanent deck-truss span serves as a falsework truss supported on barges and is shown carrying the 470 ft (143 m) anchor arm of the through-cantilever truss. This span is floated to its permanent position, then landed onto its piers by ballasting the barges. (a) Float leaves erection dock, and (b) float arrives at permanent position. (Courtesy of Bethlehem Steel Corporation.)

FIGURE 1.31 Floating-in erection of a truss span, first Chesapeake Bay Bridge, Maryland, 1952. A 480 ft (146 m) truss span, weighting 850 tons, supported on falsework consisting of a permanent deck-truss span along with temporary members, is being floated-in for landing onto its piers. Suspension bridge cables are under construction in background. (Courtesy of Bethlehem Steel Corporation.)

FIGURE 1.32 Erection of a truss span by hoisting, first Chesapeake Bay Bridge, Maryland, 1952. A 360 ft (110 m) truss span is floated into position on barges and picked clear using four sets of lifting falls. Suspension bridge deck is under construction at right. (Courtesy of Bethlehem Steel Corporation.)

FIGURE 1.33 Erection of suspension bridge deck structure, first Chesapeake Bay Bridge, Maryland, 1952. A typical four-panel through-truss deck section, weighing 99 tons, has been picked from the barge and is being raised into position using four sets of lifting falls attached to main suspension cables. Closing deck section is on barge, ready to go up next. (Courtesy of Bethlehem Steel Corporation.)

FIGURE 1.34 Greater New Orleans cantilever bridge, Louisiana, 1957. Tall double-boom deck travelers started at ends of main bridge and erected anchor spans on falsework, then the 1575 ft (480 m) main span by cantilevering to midspan. (Courtesy of Bethlehem Steel Corporation.)

FIGURE 1.35 Tower erection, second Delaware Memorial Bridge, Wilmington, Del., 1966. Tower erection traveler has reached topmost erecting position and swings into place 23-ton closing top-strut section. Tower legs were jacked apart about 2 in. (50 mm) to provide entering clearance. Traveler jumping beams are in topmost working position, above cable saddles. Tower steelwork is about 418 ft (127 m) high. Cable anchorage pier is under construction at right. First Delaware Memorial Bridge (1951) is at left. The main span of both bridges is 2150 ft (655 m). (Courtesy of Bethlehem Steel Corporation.)

FIGURE 1.36 Erecting orthotropic-plate decking panel, Poplar Street Bridge, St. Louis, Mo., 1967. A five-span, 2165 ft (660 m) continuous box-girder bridge, main span 600 ft (183 m). Projecting box ribs are 5-1/2 × 17 ft (1.7 × 5.2 m) in cross-section, and decking section is 27 × 50 ft (8.2 × 15.2 m). Decking sections were field welded, while all other connections were field bolted. Box girders are cantilevered to falsework bents using overhead "positioning travelers" (triangular structure just visible above deck at left) for intermediate support. (Courtesy of Bethlehem Steel Corporation.)

FIGURE 1.37 Erection of parallel-wire-stand (PWS) cables, Newport Bridge suspension spans, Narragansett Bay, R.I., 1968. Bridge engineering history was made at Newport with the development and application of shop-fabricated parallel-wire socketed strands for suspension bridge cables. Each Newport cable was formed of seventy-six 61-wire PWS, each 4512 ft (1375 m) long and weighing 15 tons. Individual wires are 0.202 in. (5.13 mm) in diameter and are zinc coated. Parallel-wire cables can be constructed of PWS faster and at lower cost than by traditional air spinning of individual wires (see Figures 1.25 and 1.26). (Courtesy of Bethlehem Steel Corporation.)

FIGURE 1.37A Aerial tramway tows PWS from west anchorage up side span, then on across other spans to east anchorage. Strands are about 1-3/4 in. (44 mm) in diameter.

FIGURE 1.37B Cable formers maintain strand alignment in cables prior to cable compaction. Each finished cable is about 15-1/4 in. (387 mm) in diameter. (Courtesy of Bethlehem Steel Corporation.)

FIGURE 1.38 Pipe-type anchorage for parallel-wire-strand (PWS) cables, Newport Bridge suspension spans, Narragansett Bay, R.I., 1967. Pipe anchorages shown will be embedded in anchorage concrete. Socketed end of each PWS is pulled down its pipe from upper end, then seated and shim-adjusted against heavy bearing plate at lower end. Pipe-type anchorage is much simpler and less costly than standard anchor-bar type used with aerial-spun parallel-wire cables (see Figure 1.25B). (Courtesy of Bethlehem Steel Corporation.)

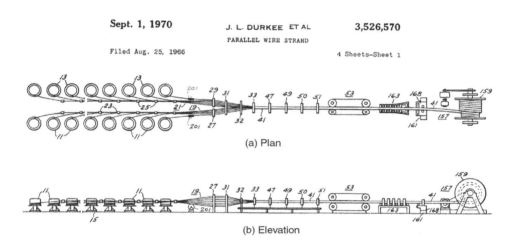

Sept. 1, 1970   J. L. DURKEE ET AL   3,526,570

PARALLEL WIRE STRAND

Filed Aug. 25, 1966   4 Sheets-Sheet 1

(a) Plan

(b) Elevation

FIGURE 1.39 Manufacturing facility for production of shop-fabricated parallel-wire strands (PWS). Prior to 1966, parallel-wire suspension bridge cables had to be constructed wire-by-wire in the field using aerial spinning procedure developed by John Roebling in the mid-19th century (refer to Figures 1.25 and 1.26). In the early 1960s a major U.S. steelwork contractor originated and developed a procedure for manufacturing and reeling parallel-wire strands, as shown in these patent drawings. A PWS can contain up to 127 wires (see Figures 1.45 and 1.46). (a) Plan view of PWS facility. Turntables 11 contain "left-hand" coils of wire and turntables 13 contain "right-hand" coils, such that wire cast is balanced in the formed strand. Fairleads 23 and 25 guide the wires into half-layplates 27 and 29, followed by full layplates 31 and 32 whose guide holes delineate the hexagonal shape of final strand 41. (b) Elevation view of PWS facility. Hexagonal die 33 contains six spring-actuated rollers that form the wires into regular-hexagon shape; and similar roller dies 47, 49, 50, and 51 maintain the wires in this shape as PWS 41 is pulled along by hexagonal dynamic clamp 53. PWS is bound manually with plastic tape at about 3 ft (1 m) intervals as it passes along between roller dies. PWS passes across roller table 163, then across traverse carriage 168, which is operated by traverse mechanism 161 to direct the PWS properly onto reel 159. Finally, reeled PWS is moved off-line for socketing. Note that wire measuring wheels 201 can be installed and used for control of strand length.

FIGURE 1.40 Suspended deck steelwork erection, Newport Bridge suspension spans, Narragansett Bay, R.I., 1968. Closing mainspan deck section is being raised into position by two cable travelers, each made up of a pair of 36 in. (0.91 m) wide-flange rolled beams that ride cables on wooden wheels. Closing section is 40-1/2 ft (12 m) long at top-chord level, 66 ft (20 m) wide and 16 ft (5 m) deep, and weighs about 140 tons. (Courtesy of Bethlehem Steel Corporation.)

FIGURE 1.41 Erection of Kansas City Southern Railway box-girder bridge, near Redland, Okla., by "launching," 1970. This nine-span continuous box-girder bridge is 2110 ft (643 m) long, with a main span of 330 ft (101 m). Box cross-section is 11 × 14.9 ft (3.35 × 4.54 m). Girders were launched in two "trains," one from north end and one from south end. A "launching nose" was used to carry leading end of each girder train up onto skidway supports as train was pushed out onto successive piers. Closure was accomplished at center of main span. (Courtesy of Bethlehem Steel Corporation.)

FIGURE 1.41A Leading end of north girder train moves across 250 ft (76 m) span 4, approaching pier 5. Main span, 330 ft (101 mm), is to right of pier 5.

FIGURE 1.41B Launching nose rides up onto pier 5 skidway units, removing girder-train leading-end sag.

FIGURE 1.41C Leading end of north girder train is now supported on pier 5.

FIGURE 1.42 Erection strengthening to withstand launching, Kansas City Southern Railway box-girder bridge, near Redland, Okla. (see Figure 1.41).

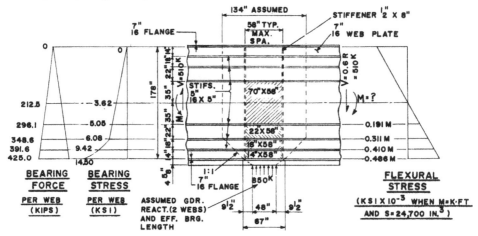

FIGURE 1.42A Typical assumed erection loading of box-girder web panels in combined moment, shear, and transverse compression.

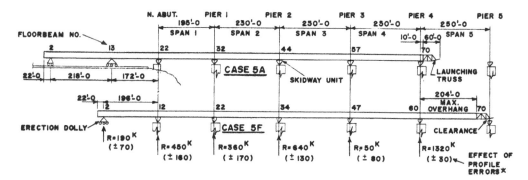

FIGURE 1.42B Launch of north girder train from pier 4 to pier 5.

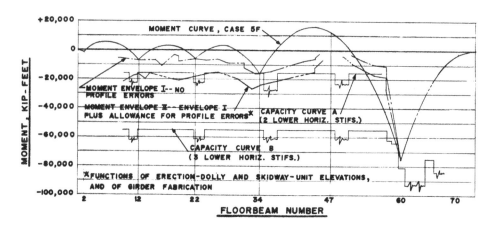

FIGURE 1.42C Negative-moment envelopes occurring simultaneously with reaction, for launch of north girder train to pier 5.

FIGURE 1.43 Erection of west arch span of twin-arch Hernando de Soto Bridge, Memphis, Tenn., 1972. The two 900 ft (274 m) continuous-truss tied-arch spans were erected by a high-tower derrick boat incorporating a pair of barges. West-arch steelwork (shown) was cantilevered to midspan over two pile-supported falsework bents. Projecting east-arch steelwork (at right) was then cantilevered to midspan (without falsework) and closed with falsework-supported other half-arch. (Courtesy of Bethlehem Steel Corporation.)

FIGURE 1.44 Closure of east side span, Commodore John Barry cantilever truss bridge, Chester, Pa., 1973. High-tower derrick boat (in background) started erection of trusses at both main piers, supported on falsework; then erected top-chord travelers for main and side spans. Sidespan traveler carried steelwork erection to closure, as shown, and falsework bent was then removed. East-mainspan traveler then cantilevered steelwork (without falsework) to midspan, concurrently with cantilever erection by west-half mainspan traveler, and trusses were closed at midspan. Commodore Barry has 1644 ft (501 m) main span, longest cantilever span in the U.S., and 822 ft (251 m) side spans. (Courtesy of Bethlehem Steel Corporation.)

FIGURE 1.45 Reel of parallel-wire strand (PWS), Akashi Kaikyo suspension bridge, Kobe, Japan, 1994. Each socketed PWS is made up of 127 0.206 in. (5.23 mm) wires, is 13,360 ft (4073 m) long, and weighs 96 tons. Plastic-tape bindings secure the strand wires at 1 m intervals. Sockets can be seen on right side of reel. These PWS are the longest and heaviest ever manufactured. (Courtesy of Nippon Steel — Kobe Steel.)

FIGURE 1.46 Parallel-wire-strand main cable, Akashi Kaikyo suspension bridge, Kobe, Japan, 1994. Main span is 6532 ft (1991 m), by far the world's longest. PWS at right is being towed across spans, supported on rollers. Completed cable is made up of 290 PWS, making a total of 36,830 wires, and has diameter of 44.2 in. (1122 mm) following compaction — largest bridge cables built to date. Each 127-wire PWS is about 2-3/8 in. (60 mm) in diameter. (Courtesy of Nippon Steel — Kobe Steel.)

FIGURE 1.47 Artist's rendering of proposed Messina Strait suspension bridge connecting Sicily with mainland Italy. The Messina Strait crossing has been under discussion since about 1870, under investigation since about 1955, and under active design since about 1975. The first realistic proposals for a crossing were made in 1969 in response to an international competition sponsored by the Italian government. There were 158 submissions — eight American, three British, three French, one German, one Swedish, and the remaining Italian. Forty of the submissions showed a single-span or multi-span suspension bridge. The enormous bridge shown has a single span of 10,827 ft (3300 m) and towers 1250 ft (380 m) high. The bridge construction problems for such a span would be tremendously challenging. (Courtesy of Stretto di Messina, S.p.A.)

References

1. Conditions of Contract and Forms of Tender, Agreement and Bond for Use in Connection with Works of Civil Engineering Construction, 6th ed. (commonly known as "The ICE Conditions of Contract"), Inst. Civil Engrs. (U.K.), 1991.
2. Copp, J.I., de Vries, K., Jameson, W.H., and Jones, J., 1945. Fabrication and Erection Controls, Rainbow Arch Bridge Over Niagara Gorge — a Symposium, *Transactions ASCE*, vol. 110.
3. Durkee, E.L., 1945. Erection of Steel Superstructure, Rainbow Arch Bridge Over Niagara Gorge — A Symposium, *Transactions ASCE*, vol. 110.
4. Durkee, J.L., 1966. Advancements in Suspension Bridge Cable Construction, *Proceedings, International Symposium on Suspension Bridges*, Laboratorio Nacional de Engenharia Civil, Lisbon.
5. Durkee, J.L., 1972. Railway Box-Girder Bridge Erected by Launching, *J. Struct. Div., ASCE*, July.
6. Durkee, J.L., and Thomaides, S.S., 1977. Erection Strength Adequacy of Long Truss Cantilevers, *J. Struct. Div., ASCE*, January.
7. Durkee, J.L., 1977. Needed: U.S. Standard Conditions for Contracting, *J. Struct. Div., ASCE*, June.
8. Durkee, J.L., 1982. Bridge Structural Innovation: A Firsthand Report, *J. Prof. Act., ASCE*, July.
9. Enquiry into the Basis of Design and Methods of Erection of Steel Box Girder Bridges. Final Report of Committee, 4 vols. (commonly known as "The Merrison Report"), HMSO (London), 1973/4.
10. Feidler, L.L., Jr., 1962. Erection of the Lewiston-Queenston Bridge, *Civil Engrg., ASCE*, November.
11. Freudenthal, A.M., Ed., 1972. The Engineering Climatology of Structural Accidents, *Proceedings of the International Conference on Structural Safety and Reliability*, Pergamon Press, Elmsford, N.Y.
12. Holgate, H., Kerry, J.G.G., and Galbraith, J., 1908. Royal Commission Quebec Bridge Inquiry Report, Sessional Paper No. 154, vols. I and II, S.E. Dawson, Ottawa, Canada.
13. Leto, I.V., 1994. Preliminary design of the Messina Strait Bridge, *Proc. Inst. Civil Engrs.* (U.K.), vol. 102(3), August.
14. Petroski, H., 1993. Predicting Disaster, *American Scientist*, vol. 81, March.
15. Pugsley, A., 1968. The Safety of Bridges, *The Structural Engineer*, U.K., July.
16. Ratay, R.T., Ed., 1996. *Handbook of Temporary Structures in Construction*, 2nd ed., McGraw-Hill, New York.
17. Schneider, C.C., 1905. The Evolution of the Practice of American Bridge Building, *Transactions ASCE*, vol. 54.
18. Sibly, P.G. and Walker, A.C., 1977. Structural Accidents and Their Causes, *Proc. Inst. Civil Engrs.* (U.K.), vol. 62(1), May.
19. Smith, D.W., 1976. Bridge Failures, *Proc. Inst. Civil Engrs.* (U.K.), vol. 60(1), August.

2

Concrete Bridge Construction

Simon A. Blank
California Department of Transportation

Michael M. Blank
U.S. Army Corps of Engineers

Luis R. Luberas
U.S. Army Corps of Engineers

2.1 Introduction

This chapter will focus on the principles and practices related to construction of concrete bridges in which construction engineering contributes greatly to the successful completion of the projects. We will first present the fundamentals of construction engineering and analyze the challenges and obstacles involved in such processes and then introduce the problems in relation to design, construction practices, project planning, scheduling and control, which are the ground of future factorial improvements in effective construction engineering in the United States. Finally, we will discuss prestressed concrete, high-performance concrete, and falsework in some detail.

2.2 Effective Construction Engineering

The construction industry is a very competitive business and many companies who engage in this marketplace develop proprietary technology in their field. In reality, most practical day-to-day issues are very common to the whole industry. Construction engineering is a combination of art and science and has a tendency to become more the art of applying science (engineering principles) and approaches to the construction operations. Construction engineering includes design, construction operation, and project management. The final product of the design team effort is to produce drawings, specifications, and special provisions for various types of bridges. A fundamental part of construction engineering is construction project management (project design, planning, scheduling, controlling, etc.).

Planning starts with analysis of the type and scope of the work to be accomplished and selection of techniques, equipment, and labor force. Scheduling includes the sequence of operations and the inter-relation of operations both at a job site and with external aspects, as well as allocation of manpower and equipment. Controlling consists of supervision, engineering inspection, detailed procedural instructions, record maintenance, and cost control. Good construction engineering analysis will produce more valuable, effective, and applicable instructions, charts, schedules, etc.

The objective is to plan, schedule, and control the construction process such that every construction worker and every activity contributes to accomplishing tasks with minimum waste of time and money and without interference. All construction engineering documents (charts, instructions, and drawings) must be clear, concise, definitive, and understandable by those who actually perform the work. As mentioned before, the bridge is the final product of design team efforts. When all phases of construction engineering are completed, this product — the bridge — is ready for service loading. In all aspects of construction engineering, especially in prestressed concrete, design must be integrated for the most effective results. The historical artificial separation of the disciplines — design and construction engineering — was set forth to take advantage of the concentration of different skills in the workplace. In today's world, the design team and construction team must be members of one team, partners with one common goal. That is the reason partnering represents a new and powerful team-building process, designed to ensure that projects become positive, ethical, and win–win experiences for all parties involved.

The highly technical nature of a prestressing operation makes it essential to perform preconstruction planning in considerable detail. Most problems associated with prestressed concrete could have been prevented by properly planning before the actual construction begins. Preconstruction planning at the beginning of projects will ensure that the structure is constructed in accordance with the plans, specifications, and special provisions, and will also help detect problems that might arise during construction. It includes (1) discussions and conferences with the contractor, (2) review of the responsibilities of other parties, and (3) familiarization with the plans, specifications, and special provisions that relate to the planned work, especially if there are any unusual conditions. The preconstruction conference might include such items as scheduling, value of engineering, grade control, safety and environmental issues, access and operational considerations, falsework requirements, sequence of concrete placement, and concrete quality control and strength requirements. Preconstruction planning has been very profitable and in many has cases resulted in substantial reduction of labor costs. More often in prestressed concrete construction, the details of tendon layout, selection of prestressing system, mild-steel details, etc., are left up to general contractors or their specialized subcontractors, with the designer showing only the final prestress and its profile and setting forth criteria. And contractors must understand the design consideration fully to select the most efficient and economical system. Such knowledge may in many cases provide a competitive edge, and construction engineering can play a very important role in it.

2.3 Construction Project Management

2.3.1 General Principles

Construction project management is a fundamental part of construction engineering. It is a feat that few, if any, individuals can accomplish alone. It may involve a highly specialized technical field or science, but it always includes human interactions, attitudes and aspects of leadership, common sense, and resourcefulness. Although no one element in construction project management will create success, failure in one of the foregoing elements will certainly be enough to promote failure and to escalate costs. Today's construction environment requires serious consultation and management of the following life-cycle elements: design (including specifications, contract clauses, and drawings), estimating, budgeting, scheduling, procurement, biddability–constructibility–operability (BCO) review, permits and licenses, site survey, assessment and layout, preconstruction and mutual understanding conference, safety, regulatory requirements, quality control (QC), construction acceptance, coordination of technical and special support, construction changes and modifications, maintenance of progress drawings (redlines), creating as-built drawings, project records, among other elements.

Many construction corporations are becoming more involved in environmental restoration either under the Resource Conservation and Recovery Act (RCRA) or under the Comprehensive Environmental Response, Compensation and Liability Act (CERCLA, otherwise commonly known as the Superfund). This new involvement requires additional methodology and considerations by managers. Some elements that would otherwise be briefly covered or completely ignored under normal considerations may be addressed and required in a site's Specific Health and Environmental Response Plan (SHERP). Some elements of the SHERP may include site health and safety staff, site hazard analysis, chemical and analytical protocol, personal protective equipment requirements and activities, instrumentation for hazard detection, medical surveillance of personnel, evacuation plans, special layout of zones (exclusion, reduction, and support), and emergency procedures.

Federal government contracting places additional demands on construction project management in terms of added requirements in the area of submittals and transmittals, contracted labor and labor standards, small disadvantaged subcontracting plans, and many other contractual certification issues. Many of these government demands are recurring elements throughout the life cycle of the project which may require adequate resource allocation (manpower) not necessary under the previous scenarios.

The intricacies of construction project management require the leadership and management skills of a unique individual who is not necessarily a specialist in any one of the aforementioned elements but who has the capacity to converse and interface with specialists in the various fields (i.e., chemists, geologists, surveyors, mechanics, etc.). An individual with a combination of an engineering undergraduate degree and a graduate business management degree is most likely to succeed in this environment. Field management experience can substitute for an advanced management degree.

It is the purpose of this section to discuss and elaborate elements of construction project management and to relate some field experiences and considerations. The information presented here will only promote further discussion and is not intended to be all-inclusive.

2.3.2   Contract Administration

Contract administration focuses on the relationships between the involved parties during the contract performance or project duration. Due to the nature of business, contract administration embraces numerous postaward and preaward functions. The basic goals of contract administration are to ensure that the owner is satisfied and all involved parties are compensated on time for their efforts. The degree and intensity of contract administration will vary from contact to contract depending upon the size and complexity of the effort to be performed. Since money is of the essence, too many resources can add costs and expenditures to the project, while insufficient resources may also cost in loss of time, in inefficiencies, and in delays. A successful construction project management program is one that has the vision and flexibility to allocate contract administrative personnel and resources wisely and that maintains a delicate balance in resources necessary to sustain required efficiencies throughout the project life cycle.

2.3.3   Project Design

Project design is the cornerstone of construction project management. In this phase, concepts are drawn, formulated, and created to satisfy a need or request. The design is normally supported by sound engineering calculations, estimates, and assumptions. Extensive reviews are performed to minimize unforeseen circumstances, avoiding construction changes or modifications to the maximum extent possible in addition to verifying facts, refining or clarifying concepts, and dismissing assumptions. This phase may be the ideal time for identification and selection of the management team.

Normally, 33, 65, 95, and 100% design reviews are standard practice. The final design review follows the 95% design review which is intended for the purpose of ensuring that review comments have been either incorporated into the design or dismissed from consideration. Reviews include design analysis reviews and BCO reviews. It can be clearly understood from the nomenclature that a BCO encompasses all facets of a project. Biddability relates to how the contact requirements are worded to ensure clarity of purpose or intent and understanding by potential construction contractors. Constructibility concentrates on how components of the work or features of the work are assembled and how they relate to the intended final product. The main purpose of the constructibility review is to answer questions, such as whether it can be built in the manner represented in the contact drawings and specifications. Interaction between mechanical, civil, electrical, and other related fields is also considered here. Operability includes aspects of maintenance and operation, warranties, services, manpower, and resource allocation during the life of the finished work.

The finished product of the design phase should include construction drawings illustrating dimensions, locations, and details of components; contract clauses and special clauses outlining specific needs of the construction contractor; specifications for mechanical, civil, and electrical or special equipment; a bidding and payment schedule with details on how parties will be compensated for work performed or equipment produced and delivered; responsibilities; and operation and maintenance (O&M) requirements. In many instances, the designer is involved throughout the construction phase for design clarification or interpretation, incorporation of construction changes or modifications to the project, and possible O&M reviews and actions. It is not uncommon to have the designer perform contract management services for the owner.

There are a number of computer software packages readily available to assist members of the management team in writing, recording, transmitting, tracking, safekeeping, and incorporating BCO comments. Accuracy of records and safekeeping of documentation regarding this process has proved to be valuable when a dispute, claim, design deficiency, or liability issue is encountered later during the project life cycle.

2.3.4 Planning and Scheduling

Planning and scheduling are ongoing tasks throughout the project until completion and occupancy by a certain date occur. Once the design is completed and the contractor selected to perform the work, the next logical step may be to schedule and conduct a preconstruction conference. Personnel representing the owner, designer, construction contractor, regulatory agencies, and any management/oversight agency should attend this conference. Among several key topics to discuss and understand, construction planning and scheduling is most likely to be the main subject of discussion. It is during this conference that the construction contractor may present how the work will be executed. The document here is considered the "baseline schedule." Thereafter, the baseline schedule becomes a living document by which progress is recorded and measured. Consequently, the baseline schedule can be updated and reviewed in a timely manner and becomes the construction progress schedule. As stated previously, the construction progress schedule is the means by which the construction contractor records progress of work, anticipates or forecasts requirements so proper procurement and allocation of resources can be achieved, and reports the construction status of work upwardly to the owner or other interested parties. In addition, the construction contractor may use progress schedule information to assist in increasing efficiencies or to formulate the basis of payment for services provided or rendered and to anticipate cash flow requirements. The construction progress schedule can be updated as needed, or mutually agreed to by the parties, but for prolonged projects it is normally produced monthly.

A dedicated scheduler, proper staffing, and adequate computer and software packages are important to accomplish this task properly. On complex projects, planning and scheduling is a full-time requirement.

2.3.5 Safety and Environmental Considerations

Construction of any bridge is a hazardous activity by nature. No person may be required to work in surroundings or under conditions that are unsafe or dangerous to his or her health. The construction project management team must initiate and maintain a safety and health program and perform a job hazard analysis with the purpose of eliminating or reducing risks of accidents, incidents, and injuries during the performance of the work. All features of work must be evaluated and assessed in order to identify potential hazards and implement necessary precautions or engineer controls to prevent accidents, incidents, and injuries.

Frequent safety inspections and continued assessment are instrumental in maintaining the safety aspects and preventive measures and considerations relating to the proposed features of work. In the safety area, it is important for the manager to be able to distinguish between accidents/incidents and injuries. Lack of recorded work-related injuries is not necessarily a measure of how safe the work environment is on the project site. The goal of every manager is to complete the job in an accident/incident- and injury-free manner, as every occurrence costs time and money.

Today's construction operational speed, government involvement, and community awareness are placing more emphasis, responsibilities, and demands on the designer and construction contractor to protect the environment and human health. Environmental impact statements, storm water management, soil erosion control plans, dust control plan, odor control measures, analytical and disposal requirements, Department of Transportation (DOT) requirements for overland shipment, activity hazard analysis, and recycling are some of the many aspects that the construction project management team can no longer ignore or set aside. As with project scheduling and planning, environmental and safety aspects of construction may require significant attention from a member of the construction management team. When not properly coordinated and executed, environmental considerations and safety requirements can delay the execution of the project and cost significant amounts of money.

2.3.6 Implementation and Operations

Construction implementation and operations is the process by which the construction project manager balances all construction and contract activities and requirements in order to accomplish the tasks. The bulk of construction implementation and operations occurs during the construction phase of the project. The construction project management team must operate in synchronization and maintain good communication channels in order to succeed in this intense and demanding phase. Many individuals in this field may contend that the implementation and operation phase of the construction starts with the site mobilization. Although it may be an indicator of actual physical activity taking place on site, construction implementation and operations may include actions and activities prior to the mobilization to the project site.

Here, a delicate balance is attempted to be maintained between all activities taking place and those activities being projected. Current activities are performed and accomplished by field personnel with close monitoring by the construction management staff. Near (approximately 1 week ahead), intermediate (approximately 2 to 4 weeks), and distant future (over 4 weeks) requirements are identified, planned, and scheduled in order to procure equipment and supplies, schedule work crews, and maintain efficiencies and progress. Coordinating progress and other meetings and conferences may take place during the implementation and operation phase.

2.3.7 Value Engineering

Some contracts include an opportunity for contractors to submit a value engineering (VE) recommendation. This recommendation is provided to either the owner or designer. The purpose of the VE is to promote or increase the value of the finished product while reducing the dollars spent or invested; in other words, to provide the desired function for minimum cost(s). VE is not intended to reduce performance, reliability, maintainability, or life expectancy below the level required to perform the basic function. Important VE evaluation criteria performed are in terms of "collateral savings" — the measurable net reductions in the owner's/agency's overall costs of construction, operations, maintenance, and/or logistics support. In most cases, collateral savings are shared between the owner/agency and the proponent of the VE by reducing the contract price or estimated cost in the amount of the instant contract savings and by providing the proponents of the VE a share of the savings by adding the amount calculated to the contract price or fee.

2.3.8   Quality Management

During the construction of a bridge, construction quality management (CQM) play a major role in quality control and assurance. CQM refers to all control measures and assurance activities instituted by the parties to achieve the quality established by the contract requirements and specifications. It encompasses all phases of the work, such as approval of submittals, procurements, storage of materials and equipment, coordination of subcontractor activities, and the inspections and the tests required to ensure that the specified materials are used and that installations are acceptable to produce the required product. The key elements of the CQM are the contractor quality control (CQC) and quality assurance (QA). To be effective, there must be a planned program of actions, and lines of authority and responsibilities must be established. CQC is primarily the construction contractor's responsibily while QA is primarily performed by an independent agency (or other than the construction contractor) on behalf of the designer or owner. In some instances, QA may be performed by the designer. In this manner, a system of checks and balances is achieved minimizing the conflicts between quality and efficiency normally developed during construction. Consequently, CQM is a combined responsibility.

In the CQC, the construction contractor is primarily responsible for (1) producing the quality product on time and in compliance with the terms of the contract; (2) verifying and checking the adequacy of the construction contractor's quality control program of the scope and character necessary to achieve the quality of construction outlined in the contract; and (3) producing and maintaining acceptable records of its QC activities. In the QA, the designated agency is primarily responsible for (1) establishing standards and QC requirements; (2) verifying and checking adequacy of the construction contractor's QC (QA for acceptance), performing special tests and inspections as required in the contract, and determining that reported deficiencies have been corrected; and (3) ensuring timely completion.

2.3.9   Partnership and Teamwork

A great deal of construction contract success, as discussed before, is attributable to partnering. Partnering should be undertaken and initiated at the earliest stage during the construction project management cycle. Some contracts may have a special clause which is intended to encourage the construction contractor to establish clear channels of communication and effective working relationships. The best approach to partnering is for the parties to volunteer to participate.

Partnering differs from the team-building concept. Team building may encourage establishing open communications and relationships when all parties share liabilities, risk, and money exposure, but not necessarily share costs of risks. The immediate goal of partnering is to establish mutual agreement(s) at the initial phases of the project on the following areas: identification of common goals, identification of common interests, establishment of lines of communication, establishment of lines of authority and decision making, commitment to cooperative problem solving, among others.

Partnering takes the elements of luck, hope, and personality out of determining project success. It facilitates workshops in which stakeholders in a specific project or program come together as a team that results in breakthrough success for all parties involved. For example, the Office of Structure Construction (OSC) of the California Department of Transportation (Caltrans) has a vision of delivery of structure construction products of the highest possible quality in partnership with their clients. And this work is not only of high quality, but is delivered in the safest, most cost-effective, and fastest manner possible. In partnership with the districts or other clients, the Office of Structure Construction (OSC) does the following to fulfill its purpose:

- Administers and inspects the construction of the Caltrans transportation structures and related facilities in a safe and efficient manner;
- Provides specialized equipment and training, standards, guidelines, and procedural manuals to ensure consistency of inspection and administration by statewide OSC staff;
- Provides consultations on safety for OSC staff and district staff performing structure construction inspection work;

- Conducts reviews and provides technical consultation and assistance for trenching and shoring temporary support and falsework construction reviews;
- Provides technical recommendations on the preparations of structure claims and the contract change orders (CCOs);
- Provides construction engineering oversight on structure work on non-state-administrated projects;
- Conducts BCO review.

2.3.10 Project Completion and Turnover of Facility

Success in construction project management may be greatly impacted during project completion and turnover of the facilities to the user or owner. The beginning of the project completion and turnover phase may be identified by one of the following: punch list developed, prefinal inspections scheduled, support areas demobilized, site restoration initiated, just to mention a few. Many of the problems encountered during this last phase may be avoided or prevented with proper user or owner participation and involvement during the previous phases, particularly during the construction where changes and modifications may have altered the original design. A good practice in preventing conflicts during the completion and turnover of the facilities is to invite the owner or user to all construction progress meetings and acceptance inspections. In that manner, the user or owner is completely integrated during the construction with ample opportunity to provide feedback and be part of the decision-making process. In addition, by active participation, the owner or user is being informed and made aware of changes, modifications, and/or problems associated with the project.

2.4 Major Construction Considerations

Concrete bridge construction involves site investigation; structure design; selection of materials — steel, concrete, aggregates, and mix design; workmanship of placment and curing of concrete; and handling and maintenance of the structure throughout its life. Actually, site investigations are made of any structure, regardless of how insignificant it may be. The site investigation is very important for intelligent design of the bridge structures and has a significant influence on selection of the material and mix. A milestone is to investigate the fitness of the location to satisfy the requirements of the bridge structure. Thus, investigation of the competence of the foundation to carry the service load safely and an investigation of the existence of forces or substances that may attack the concrete structure can proceed. Of course, the distress or failure may have several contributing causal factors: unsuitable materials, construction methods, loading conditions, faulty mix design, design mistakes, conditions of exposure, curing condition, or environmental factors.

2.5 Structural Materials

2.5.1 Normal Concrete

Important Properties

Concrete is the only material that can be made on site, and is practically the most dependable and versatile construction material used in bridge construction. Good durable concrete is quality concrete that meets all structural and aesthetic requirements for a period of structure life at minimum cost. We are looking for such properties as workability in the fresh condition; strength in accordance with design, specifications, and special provisions; durability; volume stability; freedom from blemishes (scaling, rock pockets, etc.); impermeability; economy; and aesthetic appearance. Concrete when properly designed and fabricated can actually be crack-free not only under normal service loads, but also under moderate overload, which is very attractive for bridges that are exposed to an especially corrosive atmosphere.

The codes and specifications usually specify the minimum required strength for various parts of a bridge structure. The required concrete strength is determined by design engineers. For cast-in-place concrete bridges, a compressive strength of 3250 to 5000 psi (22 to 33 MPa) is usual. For precast structure compressive strength of 4000 to 6000 psi (27 to 40 MPa) is often used. For special precast, prestressed structures compressive strength of 6000 to 8000 psi (40 to 56 MPa) is used. Other properties of concrete are related to the strength, although not necessarily dependent on the strength.

Workability is the most important property of fresh concrete and depends on the properties and proportioning of the materials: fine and coarse aggregates, cement, water, and admixtures. Consistency, cohesiveness, and plasticity are elements of workability. Consistency is related to the fluidity of mix. Just adding water to a batch of concrete will make the concrete more fluid or "wetter," but the quality of the concrete will diminish. Consistency increases when water is added and an average of 3% in total water per batch will change the slump about 1 in. (2.54 cm). The research and practice show that workability is a maximum in concrete of medium consistency, between 3 in. (7.62 cm) and 6 in. (15.24 cm) slump. Very dry or wet mixes produce less-workable concrete. Use of relatively harsh and dry mixes is allowed in structures with large cross sections, but congested areas containing much reinforcement steel and embedded items require mixes with a high degree of workability.

A good and plastic mixture is neither harsh nor sticky and will not segregate easily. Cohesiveness is not a function of slump, as very wet (high-slump) concrete lacks plasticity. On the other hand, a low-slump mix can have a high degree of plasticity. A harsh concrete lacks plasticity and cohesiveness and segregates easily.

Workability has a great effect on the cost of placing concrete. Unworkable concrete not only requires more labor and effort in placing but also produces rock pockets and sand streaks, especially in small congested forms. It is a misconception that compaction or consolidation of concrete in the form can be done with minimum effort if concrete is fluid or liquid to flow into place. It is obvious that such concrete will flow in place but segregate badly, so that large aggregate will settle out of the mortar and excess water will rise to the top surface. And unfortunately, this error in workmanship will become apparent after days, even months later, showing up as cracks, low strength, and general inferiority of concrete. The use of high-range water-reducing admixtures (superplasticizers) allows placing of high-slump, self-leveling concrete. They increase the strength of concrete and provide great workability without adding an excessive amount of water. As an example of such products used in the Caltrans is PolyHeed 997 which meets the requirements for a Type A, water-reducing admixture specified in ASTM C 494-92, Corps of Engineers CRD-C 87-93, and AASHTO M 194-87, the Standard Specifications for chemical admixtures for concrete.

Special Consideration for Cold-Weather Construction

Cold weather can damage a concrete structure by freezing of fresh concrete before the cement has achieved final set and by repeated cycles of freezing of consequent expansion of water in pores and openings in hardened concrete. Causes of poor frost resistance include poor design of construction joints, segregation of concrete during placement; leaky formwork; poor workmanship, resulting in honeycomb and sand streaks; and insufficient or absent drainage, permitting water to accumulate against concrete. In order to provide resistance against frost adequate drainage should be designed. If horizontal construction joints are necessary, they should be located below the low-water or above the high-water line about 2 to 3 ft (0.6 to 1 m). Previously placed concrete must be cleaned up completely. Concrete mix should have a 7% (max) air for ½ in. (12.7 mm) or ¾ in. (19 mm) (max) aggregate, ranging down to 3 to 4% for cobble mixes. It is essential to use structurally sound aggregates with low porosity. The objective of frost-resistant concrete mix is to produce good concrete with smooth, dense, and impermeable surface. This can be implemented by good construction techniques used in careful placement of concrete as near as possible to its final resting place, avoiding segregation, sand streaks, and honeycomb under proper supervision, quality control, and assurance.

Sudden changes in temperature can stress concrete and cause cracking or crazing. A similar condition exists when cold water is applied to freshly stripped warm concrete, particularly during hot weather. For the best results, the temperature difference should not exceed 25°F between concrete and curing water.

In cases when anchor bolt holes were left exposed to weather and filled with water, freezing of water exerted sufficient force to crack concrete. This may happen on the bridge pier cap under construction.

Concrete Reinforcement and Placement

The optimum conditions for structural use is a medium slump of concrete and compaction by vibrators. A good concrete with low slump for the placing conditions can be ruined by insufficient or improper consolidation. Even workable concrete may not satisfy the needs of the bridge structure if it is not properly consolidated, preferably by vibration. An abrupt change in size and congestion of reinforcement not only makes proper placing of concrete difficult but also causes cracks to develop. Misplacement of reinforcement within concrete will greatly contribute to development of structural cracks. The distress and failure of concrete are mostly caused by ignorance, carelessness, wrong assumptions, etc.

Concrete Mix and Trial Batches

The objective of concrete mix designs and trial batches is to produce cost-effective concrete with sufficient workability, strength, durability, and impermeability to meet the conditions of placing, finishing characteristics, exposure, loading, and other requirements of bridge structures. A complete discussion of concrete mixes and materials can be found in many texts such as *Concrete Manual* by Waddel [1]. The purpose of trial batches is to determine strength, water–cement ratio, combined grading of aggregates, slump, type and proportioning of cement, aggregates, entrained air, and admixtures as well as scheduling of trial batches and uniformity. Trial batches should always be made for bridge structures, especially for large and important ones. They should also be made in cases where there is no adequate information available for existing materials used in concrete mixes, and they are subjected to revision in the field as conditions require.

Consideration to Exposure Condition

Protection of waterfront structures should be considered when they are being designed. Designers often carefully consider structural and aesthetic aspects without consideration of exposure conditions. Chemical attack is aggravated in the presence of water, especially in transporting the chemiclas into the concrete through cracks, honeycombs, or pores in surfaces. Use of chamfers and fillers is good construction practice. Chamfering helps prevent spalling and chipping from moving objects. Fillets in reentrant corners eliminate possible scours or cracking. Reinforcement should be well covered with sound concrete and in most cases the 3 in. (7.62 cm) coverage is specified. First-class nonreactive and well-graded aggregates in accordance with the UBC standard should be used. Cement Type II or Type Y with a low of C_3 should be used. Careful consideration should be given to the use of an approved pozzolan with a record of successful usage in a similar exposure. Mix design should contain an adequate amount of entrained air and other parameters in accordance with specifications or a special provision for a particular project. The concrete should be workable with slump and water–cement ratio as low as possible and containing at least 560 pcy (332 kg/m^3). To reduce mixing water for the same workability and, by the same token, to enhance strength and durability, a water-reducing admixture is preferred. The use of calcium chloride and Type III cement for acceleration of hardening and strength development is precluded. Concrete should be handled and placed with special care to avoid segregation and prevent honeycomb and sand streaks. The proper cure should be taken for at least seven days before exposure.

2.5.2 High-Performance Concrete

High-performance concrete (HPC) is composed of the same materials used in normal concrete, but proportioned and mixed to yield a stronger, more durable product. HPC structures last much longer and suffer less damage from heavy traffic and climatic condition than those made with conventional concrete. To promote the use of HPC in highway structures in the United States, a group of concrete experts representing the state DOTs, academia, the highway industry, and the Federal Highway Administration (FHWA) has developed a working definition of HPC, which includes performance criteria and the standard tests to evaluate performance when specifying an HPC mixture. The designer determines what level of strength, creep, shrinkage, elasticity, freeze/thaw durability, abrasion resistance, scaling resistance, and chloride permeability are needed. The definition specifies what tests grade of HPC satisfies those requirements and what tests to perform to confirm that the concrete meets that grade.

An example of the mix design for the 12,000-psi high-strength concrete used in the Orange County courthouse in Florida follows:

Gradient	Weight (pounds)
Cement, Type 1	900
Fly ash, Class F	72
Silica fume	62
Natural sand	980
No. 8 granite aggregate	1,780
Water	250
Water reducer	2 oz per cubic hundredweight
Superplasticizer	35 oz per cubic hundredweight

The Virginia and Texas DOTs have already started using HPC that is ultra-high-strength concrete 12,000 to 15,000 psi (80 to 100 MPa) in bridge construction and rehabilitation of the existing bridges [2].

2.5.3   Steel

All reinforcing steel for bridges is required to conform to specifications of ASTM Designation A615, Grade 60 or low-alloy steel deformed bars conforming to ASTM Designation A706. Prestressing steel: high-tensile wire conforming to ASTM Designations: A421, including Supplement I, High-tensile wire strand A416, Uncoated high-strength steel bars: A722, are usually used. All prestressing steel needs to be protected against physical damage and rust or other results of corrosion at all times from manufacture to grouting or encasing in concrete. Prestressing steel that has physical damage at any time needs to be rejected. Prestressing steel for post-tensioning that is installed in members prior to placing and curing of the concrete needs to be continuously protected against rust or other corrosion until grouted, by means of a corrosion inhibitor placed in the ducts or applied to the steel in the duct.

The corrosion inhibitor should conform to the specified requirements. When steam curing is used, prestressing steel for post-tensioning should not be installed until the stem curing is completed. All water used for flushing ducts should contain either quick lime (calcium oxide) or slaked lime (calcium hydroxide) in the amount of 0.01 kg/l. All compressed air used to blow out ducts should be oil free.

2.6   Construction Operations

2.6.1   Prestressing Methods

If steel reinforcement in reinforced concrete structures is tensioned against the concrete, the structure becomes a prestressed concrete structure. This can be accomplished by using pretensioning and post-tensioning methods.

Pretensioning

Pretensioning is accomplished by stressing tendons, steel wires, or strands to a predetermined amount. While stress is maintained in the tendons, concrete is placed in the structure. After the concrete in the structure has hardened, the tendons are released and the concrete bonded to the tendons becomes prestressed.

Widely used in pretensioning techniques are hydraulic jacks and strands composed of several wires twisted around a straight center wire. Pretensioning is a major method used in the manufacture of prestressed concrete in the United States. The basic principles and some of the methods currently used in the United States were imported from Europe, but much has been done in the United States to

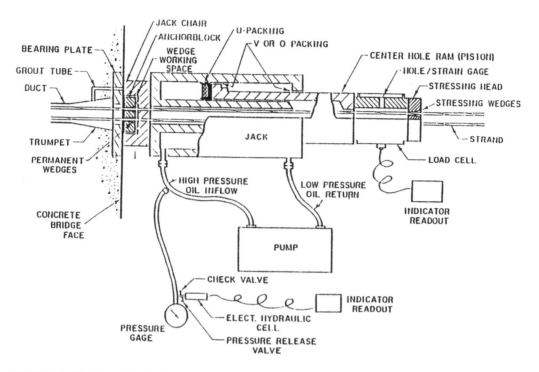

FIGURE 2.1 Typical post-tensioning system.

develop and adapt manufacturing procedures. One such adaptation employs pretensioned tendons which do not pass straight through the concrete member, but are deflected or draped into a trajectory that approximates a curve. This method is very widely practiced in the fabrication of precast bridge girders in the United States.

Post-Tensioning

A member is called as post-tensioned when the tendons are tensioned after the concrete has hardened and attained sufficient strength (usually 70% final strength) to withstand the prestressing force, and each end of the tendons are anchored. Figure 2.1 shows a typical post-tensioning system. A common method used in the United States to prevent tendons from bonding to the concrete during placing and curing of the concrete is to encase the tendon in a mortar-tight metal tube or flexible metal hose before placing it in the forms. The metal hose or tube is referred to as a sheath or duct and remains in the structure. After the tendons have been stressed, the void between the tendons and the duct is filled with grout. The tendons become bonded to the structural concrete and protected from corrosion [3]. Construction engineers can utilize prestressing very effectively to overcome excessive temporary stresses or deflections during construction, for example, using cantilevering techniques in lieu of falsework.

Prestressing is not a fixed state of stress and deformation, but is time dependent. Both concrete and steel may be deformed inelastically under continued stress. After being precompressed, concrete continues to shorten with time (creep). Loss of moisture with time also contributes to a shortening (shrinkage). In order to reduce prestress losses due to creep and shrinkage and to increase the level of precompression, use of not only higher-strength steel but also higher-strength concrete that has low creep, shrinkage, and thermal response is recommended. New chemical admixtures such as high-range water-reducing admixtures (superplasticizers) and slag are used for producing high-performance concrete and for ultra-high-strength concrete. The new developments are targeted to producing high-strength steel that is "stabilized" against stress relaxation which leads to a reduction of stress in tendons, thus reducing the prestress in concrete.

2.6.2 Fabrication and Erection Stages

During construction, not all elements of a bridge have the same stresses they were designed for. That is the reason it is a very important part of construction engineering to be aware of this and to make sure that appropriate steps have been taken. For example, additional reinforcement will be added to the members in the fabrication stage and delivered to the job site for erection.

In the case of cast-in-place box-girder bridge construction the sequences of prestressing tendons have to be engineered step-by-step to ensure that the structure will have all parameters for future service load after completion of this stage.

The sequence of the erection itself may produce additional stresses that structures or portions of the structures were not designed for. These stresses and the stability of structures during erection are a big concern that is often overlooked by designers and contractors — construction sequences play a very important role in the erection of a segmental type of bridge. It seems that we have to give more attention to analysis of the role of the construction engineering implementation of such erections. And, yes, sometimes the importance of construction engineering to accomplish safe and efficient fabrication and erection of bridge structures (precast, prestressed girders, cast-in-pile) is not sufficiently emphasized by design engineers and/or fabrication, erection contractors.

Unfortunately, we have to admit that the design set of drawings even for an important bridge does not include the erection scheme. And, of course, we can show many examples of misplaced erection efforts on the part of the designer, but our goal is to show why it happened and to make efforts to pay more attention to the fabrication and erection stages. Even if such an erection scheme is included in the design drawings, contractors are not supposed to rely solely on what is provided by the designer's erection plan.

Sometimes a design can be impractical, or it may not be suitable in terms of the erection contractor's equipment and experience. Because the erection plans usually are very generalized and because not enough emphasis is given to the importance of this stage, it is important that the designer understand the contractor's proposed method so that the designer can determine if these methods are compatible with the plans, specifications, and requirements of the contract. This is the time that any differences should be resolved. The designer should also discuss any contingency plan in case the contractor has problems. In many instances, the designer is involved throughout the construction phase for design and specification clarification or interpretation, incorporation of construction changes or modifications to the project, and possible O&M reviews/action.

2.6.3 Construction of Segmental Bridges

The first precast segmental box-girder bridge was built by Jean Muller, the Choisy-le-Roi-Bridge crossing the Seine River in 1962. In North America (Canada), a cast-in-place segmental bridge on the Laurentian Autoroute, near Ste. Adele, Quebec, in 1964 and a precast segmental bridge crossing the Lievre River near Notre Dame du Laus also in Quebec in 1967 were constructed. In the United States, the first precast segmental bridge was completed in 1973 in Corpus Christi (Texas). The Pine Valley Creek Bridge with five spans (270 + 340 + 450 + 380 + 270 ft) supported by 340-ft-high pier as shown in Figures 2.2 through 2.5 is the first cast-in-place segmental bridge constructed in the United States in 1974 using the cantilever method. The ends of the bridge are skewed to fit the bridge into the canyon. The superstructure consists of two parallel box structures each providing a roadway width of 40 ft between railings. The superstructures are separated by a 38-ft median.

Segmental cantilever construction is a fairly recent development, and the concept has been improved and used successfully to build bridges throughout the world. Its unique characteristic of needing no ground-supported falsework makes the method attractive for use over congested streets, waterways, deep gorges, or ocean inlets. It has been used for spans of less than 100 ft, all the way to the current record span of 755 ft over the Urato River in Japan. Another advantage of the method lies in its economy and efficiency of material use. Construction of segmental bridges can be classified by three methods: balanced cantilever, span-by-span, and progressive placement or incremental launching.

FIGURE 2.2   Pine Valley Creek Bridge — construction at Pier 4.

FIGURE 2.3   Pine Valley Creek Bridge — pier construction.

FIGURE 2.4   Pine Valley Creek Bridge — girder construction.

FIGURE 2.5   Pine Valley Creek Bridge — construction completion.

2.6.4 Construction of HPC Bridges

The first U.S. bridge was built with HPC under the Strategic Highway Research Program (SHRP) in Texas in 1996. The FHWA and the Texas DOT in cooperation with the Center for Transportation Research (CTR) at the University of Texas at Austin sponsored a workshop to showcase HPC for bridges in Houston in 1996. The purpose of the event was to introduce the new guidelines to construction professionals and design engineers, and to show how HPC was being used to build more durable structures. It also focused on the pros and cons of using HPC, mix proportioning, structural design, HPC in precast prestressed and cast-in-place members, long-term performance, and HPC projects in Nebraska, New Hampshire, and Virginia. The showcase had a distinctly regional emphasis because local differences in cements, aggregates, and prestressing fabricators have a considerable impact on the design and construction of concrete structures. In Texas, concrete can be produced with compressive strength of 13,000 to 15,000 psi (900 to 1000 MPa).

The first bridge to utilize HPC fully in all aspects of design and construction is the Louetta Road Overpass on State Highway 249 in Houston. The project consists of two U-beam bridges carrying two adjacent lanes of traffic. The spans range from 121.5 to 135.5 ft (37 to 41.3 m) long. The HPC is about twice as strong as conventional concrete. It costs an average of $260/m² ($24/ft²) of deck area, a price compatible with the 12 conventional concrete bridges on the same project. The second Texas HPC bridge located in San Angelo carries the eastbound lanes of U.S. Route 67 over the North Concho River, U.S. 87, and the South Orient railroad. The 954-ft (291-m) HPC I-beam bridge runs parallel to a conventional concrete bridge. The HPC was chosen for the eastbound lanes because the span crossing the North Concho River was 157 ft (48 m) long. This distance exceeds the capacity of Texas conventional prestressed concrete U-beam simple-span construction. The San Angelo Bridge presents an ideal opportunity for comparing HPC and conventional concrete. The first spans of two bridges are the same length and width, making it easy to compare the cost and performance between HPC and conventional concrete. The comparison indicated that conventional concrete lanes of the first span required seven beams with 5.6 ft (1.7 m) spacing, while the HPC span required only four beams with 11 ft (3.4 m) spacing.

The Louetta Road Overpass using HPC is expected to have a useful life of 75 to 100 years, roughly double the average life of a standard bridge. A longer life span means not only lower user cost, but motorists will encounter fewer lane closures and other delays caused by maintenance work. At the present time 15 HPC bridges have been built in the United States.

2.7 Falsework

Falsework may be defined as a temporary framework on which the permanent structure is supported during its construction. The term *falsework* is universally associated with the construction of cast-in-place concrete structures, particularly bridge superstructures. The falsework provides a stable platform upon which the forms may be built and furnish support for the bridge superstructure.

Falsework is used in both building and bridge construction. The temporary supports used in building work are commonly referred to as "shoring." It is also important to note the difference between "formwork" and "falsework." Formwork is used to retain plastic concrete in its desired shape until it has hardened. It is designed to resist the fluid pressure of plastic concrete and additional pressure generated by vibrators. Because formwork does not carry dead load of concrete, it can be removed as soon as the concrete hardens. Falsework does carry the dead load of concrete, and therefore it has to remain in place until the concrete becomes self-supporting. Plywood panels on the underside of a concrete slab serve both as a formwork and as a falsework member. For design, however, such panels are considered to be forms in order to meet all design and specification requirements applied to them.

Bridge falsework can be classified in two types: (1) conventional systems (Figure 2.6) in which the various components (beams, posts, caps, bracings, etc.) are erected individually to form the completed system and (2) proprietary systems in which metal components are assembled into modular units that

FIGURE 2.6 Falsework at I-80 HOV construction, Richmond, CA.

can be stacked, one above the other, to form a series of towers that compose the vertical load-carrying members of the system.

The contractor is responsible for designing and constructing safe and adequate falsework that provides all necessary rigidity, supports all load composed, and produces the final product (structure) according to the design plans, specifications, and special provisions. It is very important also to keep in mind that approval by the owner of falsework working drawings or falsework inspection will in no way relieve the contractor of full responsibility for the falsework. In the state of California, any falsework height that exceeds 13 ft (4 m) or any individual falsework clear span that exceeds 17 ft (5 m) or where provision for vehicular, pedestrian, or railroad traffic through the falsework is made, the drawings have to be signed by the registered civil engineer in the state of California. The design drawings should include details of the falsework removal operations, methods and sequences of removal, and equipment to be used. The drawings must show the size of all load-supporting members, connections and joints, and bracing systems. For box-girder structures, the drawings must show members supporting sloping exterior girders, deck overhangs, and any attached construction walkway. All design-controlling dimensions, including beam length and spacing, post locations and spacing, overall height of falsework bents, and vertical distance between connectors in diagonal bracing must be shown.

It is important that falsework construction substantially conform to the falsework drawings. As a policy consideration, minor deviations to suit field conditions or the substitution of materials will be permitted if it is evident by inspection that the change does not increase the stresses or deflections of any falsework members beyond the allowable values, nor reduce the load-carrying capacity of the overall falsework system. If revision is required, the approval of revised drawings by the state engineer is also required. Any change in the approved falsework design, however minor it may appear to be, has the potential to affect adversely the structural integrity of the falsework system. Therefore, before approving any changes, the engineer has to be sure that such changes will not affect the falsework system as a whole.

References

1. Waddel, J. J., *Concrete Manual*, 1994.
2. *Focus Mag.*, May 1996.
3. Libby, J., *Modern Prestressed Concrete*, 1994.
4. Gerwick, B.C., Jr., *Construction of Prestressed Concrete Structures*, 1994.

5. Fisk, E.R., *Construction Project Administration*, John Wiley & Sons, New York, 1994.
6. Blank, M.M., *Selected Published Articles from 1986 to 1995*, Naval Air Warfare Center, Warmminster, PA, 1996.
7. Blank, M.M. and Blank, S.A., *Effective Construction Management Tools*, Management Division, Naval Air Warfare Center, Warmminster, PA, 1995.
8. Godfrey, K.A., *Partnering in Design and Construction*, McGraw-Hill, New York, 1995.
9. Blank, M.M. et al., Partnering: the key to success, *Found. Drilling*, March/April, 1997.
10. Kubai, M.T. *Engineering Quality in Construction Partnering and TQM*, McGraw-Hill, New York, 1994.
11. Rubin, D.K., A burning sensation in Texas, *Eng. News Rec.*, July 12, 1993.
12. Partnering Guide for Environmental Missions of the Air Force, Army and Navy, Prepared by a Tri-Service Committee, Air Force, Army, and Navy, July 1996.
13. Post, R.G., Effective partnering, *Construction*, June 1996.
14. Schriener, J., Partnering, TQM, ADF, low insurance cost, *Eng. News Rec.*, January 15, 1995.
15. Caltrans, *Standard Specifications*, California Department of Transportation, Sacramento, 1997.
16. Caltrans, *A Guide for Field Inspection of Cast-in-Place Post-Tensioned Structures*, California Department of Transportation, Sacramento, 1992.

3

Substructures of Major Overwater Bridges

Ben C. Gerwick, Jr.
*Ben C. Gerwick Inc. and University
of California, Berkeley*

3.1 Introduction

The design and construction of the piers for overwater bridges present a series of demanding criteria. In service, the pier must be able to support the dead and live loads successfully, while resisting environmental forces such as current, wind, wave, sea ice, and unbalanced soil loads, sometimes even including downslope rock fall. Earthquake loadings present a major challenge to design, with cyclic reversing motions propagated up through the soil and the pier to excite the superstructure. Accidental forces must also be resisted. Collision by barges and ships is becoming an increasingly serious hazard for bridge piers in waterways, both those piers flanking the channel and those of approaches wherever the water depth is sufficient.

Soil–structure foundation interaction controls the design for dynamic and impact forces. The interaction with the superstructure is determined by the flexibility of the entire structural system and its surrounding soil.

Rigid systems attract very high forces: under earthquake, the design forces may reach 1.0 *g*, whereas flexible structures, developing much less force at longer periods, are subject to greater deflection drift. The design must endeavor to obtain an optimal balance between these two responses. The potential for scour due to currents, amplified by vortices, must be considered and preventive measures instituted.

Constructibility is of great importance, in many cases determining the feasibility. During construction, the temporary and permanent structures are subject to the same environmental and accidental loadings as the permanent pier, although for a shorter period of exposure and, in most cases, limited to a favorable time of the year, the so-called weather window. The construction processes employed must therefore be practicable of attainment and completion. Tolerances must be a suitable compromise between practicability and future performance. Methods adopted must not diminish the future interactive behavior of the soil–structure system.

The design loadings for overwater piers are generally divided into two limit states, one being the limit state for those loadings of high probability of occurrence, for which the response should be essentially elastic. Durability needs to be considered in this limit state, primarily with respect to corrosion of exposed and embedded steel. Fatigue is not normally a factor for the pier concepts usually considered, although it does enter into the considerations for supplementary elements such as fender systems and temporary structures such as dolphins if they will be utilized under conditions of cyclic loading such as waves. In seismic areas, moderate-level earthquakes, e.g., those with a return period of 300 to 500 years, also need to be considered.

The second limit state is that of low-probability events, often termed the "safety" or "extreme" limit state. This should include the earthquake of long return period (1000 to 3000 years) and ship collision by a major vessel. For these, a ductile response is generally acceptable, extending the behavior of the structural elements into the plastic range. Deformability is essential to absorb these high-energy loads, so some damage may be suffered, with the provision that collapse and loss of life are prevented and, usually, that the bridge can be restored to service within a reasonable time.

Plastic hinging has been adopted as a principle for this limit state on many modern structures, designed so that the plastic hinging will occur at a known location where it can be most easily inspected and repaired. Redundant load paths are desirable: these are usually only practicable by the use of multiple piles.

Bridge piers for overwater bridges typically represent 30 to 40% of the overall cost of the bridge. In cases of deep water, they may even reach above 50%. Therefore, they deserve a thorough design effort to attain the optimum concept and details.

Construction of overwater bridge piers has an unfortunate history of delays, accidents, and even catastrophes. Many construction claims and overruns in cost and time relate to the construction of the piers. Constructibility is thus a primary consideration.

The most common types of piers and their construction are described in the following sections.

3.2 Large-Diameter Tubular Piles

3.2.1 Description

Construction of steel platforms for offshore petroleum production as well as deep-water terminals for very large vessels carrying crude oil, iron, and coal, required the development of piling with high axial and lateral capacities, which could be installed in a wide variety of soils, from soft sediments to rock. Lateral forces from waves, currents, floating ice, and earthquake as well as from berthing dominated the design. Only large-diameter steel tubular piles have proved able to meet these criteria (Figures 3.1 and 3.2).

Such large piling, ranging from 1 to 3 m in diameter and up to over 100 m in length required the concurrent development of very high energy pile-driving hammers, an order of magnitude higher than those previously available. Drilling equipment, powerful enough to drill large-diameter sockets in bedrock, was also developed (Figure 3.3).

Thus when bridge piers were required in deeper water, with deep sediments of varying degrees or, alternatively, bare rock, and where ductile response to the lateral forces associated with earthquake, ice,

FIGURE 3.1 Large-diameter steel tubular pile, Jamuna River Bridge, Bangladesh.

FIGURE 3.2 Driving large-diameter steel tubular pile.

FIGURE 3.3 Steel tubular pile being installed from jack-up barge. Socket will be drilled into rock and entire pile filled with tremie concrete.

and ship impact became of equal or greater importance than support of axial loads, it was only natural that technology from the offshore platform industry moved to the bridge field.

The results of this "lateral" transfer exceeded expectations in that it made it practicable and economical to build piers in deep waters and deep sediments, where previously only highly expensive and time-consuming solutions were available.

3.2.2 Offshore Structure Practice

The design and construction practices generally follow the Recommended Practice for Planning, Designing and Constructing Fixed Offshore Platforms published by the American Petroleum Institute, API-RP2A [1]. This recommended practice is revised frequently, so the latest edition should always be used. Reference [2] presents the design and construction from the construction contractor's point of view.

There are many variables that affect the designs of steel tubular piles: diameter, wall thickness (which may vary over the length), penetration, tip details, pile head details, spacing, number of piles, geometry, and steel properties. There must be consideration of the installation method and its effect on the soil–pile interaction. In special cases, the tubular piles may be inclined, i.e., "raked" on an angle from vertical.

In offshore practice, the piles are almost never filled with concrete, whereas for bridge piers, the designer's unwillingness to rely solely on skin friction for support over a 100-year life as well as concern for corrosion has led to the practice of cleaning out and filling with reinforced concrete. A recent advance has been to utilize the steel shell along with the concrete infill in composite action to increase strength and stiffness. The concrete infill is also utilized to resist local buckling under overload and extreme conditions. Recent practice is to fill concrete in zones of high moment.

Tubular piles are used to transfer the superimposed axial and lateral loads and moments to the soil. Under earthquake, the soil imparts dynamic motions to the pile and hence to the structure. These interactions are highly nonlinear. To make matters even more complex, the soils are typically nonuniform throughout their depth and have different values of strength and modulus.

In design, axial loads control the penetration while lateral load transfer to the soil determines the pile diameter. Combined pile stresses and installation stresses determine the wall thickness. The interaction of the pile with the soil is determined by the pile stiffness and diameter. These latter lead to the development of a P–y curve, P being the lateral shear at the head of the pile and y being the deflection along the pile. Although the actual behavior is very complex and can only be adequately solved by a computerized final design, an initial approximation of three diameters can give an assumed "point of fixity" about which the top of the pile bends.

Experience and laboratory tests show that the deflection profile of a typical pile in soft sediments has a first point of zero deflection about three diameters below the mudline, followed by deflection in reverse bending and finally a second point of zero displacement. Piles driven to a tip elevation at or below this second point have been generally found to develop a stable behavior in lateral displacement even under multiple cycles of high loading.

If the deflection under extreme load is significant, P–Δ effects must also be considered. Bridge piers must not only have adequate ultimate strength to resist extreme lateral loads but must limit the displacement to acceptable values. If the displacement is too great, the P–Δ effect will cause large additional bending moments in the pile and consequently additional deflection.

The axial compressive behavior of piles in bridge piers is of dominant importance. Settlement of the pile under service and extreme loads must be limited. The compressive axial load is resisted by skin friction along the periphery of the pile, by end bearing under the steel pile tip, and by the end bearing of the soil plug in the pile tip. This latter must not exceed the skin friction of the soil on the inside of the pile, since otherwise the plug will slide upward. The actual characteristics of the soil plug are greatly affected by the installation procedures, and will be discussed in detail later.

Axial tension due to uplift under extreme loads such as earthquakes is resisted by skin friction on the periphery and the deadweight of the pile and footing block.

Pile group action usually differs from the summation of individual piles and is influenced by the stiffness of the footing block as well as by the applied bending moments and shears. This group action and its interaction with the soil are important in the final design, especially for dynamic loading such as earthquakes.

API-RP2A Section G gives a design procedure for driven steel tubular piles as well as for drilled and grouted piles.

Corrosion and abrasion must be considered in determining the pile wall thickness. Corrosion typically is most severe from just below the waterline to just above the wave splash level at high tide, although another vulnerable location is at the mudline due to the oxygen gradient. Abrasion typically is most severe at the mudline because of moving sands, although suspended silt may cause abrasion throughout the water column.

Considering a design lifetime of a major bridge of 100 years or more, coatings are appropriate in the splash zone and above, while sacrificial anodes may be used in the water column and at the mudline. Additional pile wall thickness may serve as sacrificial steel: for seawater environment, 10 to 12 mm is often added.

3.2.3 Steel Pile Design and Fabrication

Tubular steel piles are typically fabricated from steel plate, rolled into "cans" with the longitudinal seam being automatically welded. These cans are then joined by circumferential welds. Obviously, these welds are critical to the successful performance of the piles. During installation by pile hammer, the welds are often stressed very highly under repeated blows: defective welds may crack in the weld or the heat-affected zone (HAZ). Welds should achieve as full joint penetration as practicable, and the external weld profile should merge smoothly with the base metal on either side.

API-RP2A, section L, gives guidance on fabrication and welding. The fabricated piles should meet the specified tolerances for both pile straightness and for cross section dimensions at the ends. These latter control average diameter and out-of-roundness. Out-of-roundness is of especial concern as it affects the ability to match adjacent sections for welding.

Inspection recommendations are given in API-RP2A, section N. Table N.4-1, with reference to structural tubulars, calls for 10% of the longitudinal seams to be verified by either ultrasonic (UT) or radiography (RT). For the circumferential weld seams and the critical intersection of the longitudinal and circumferential seams, 100% UT or RT is required.

Because of the typically high stresses to which piles supporting bridge piers are subjected, both under extreme loads and during installation, as well as the need for weldability of relatively thick plates, it is common to use a fine-grained steel of 290 to 350 MPa yield strength for the tubular piles.

Pile wall thickness is determined by a number of factors. The thickness may be varied along the length, being controlled at any specific location by the loading conditions during service and during installation.

The typical pile used for a bridge pier is fixed at the head. Hence, the maxima combined bending and axial loads will occur within the 1½ diameters immediately below the bottom of the footing. Local buckling may occur. Repeated reversals of bending under earthquake may even lead to fracture. This area is therefore generally made of thicker steel plate. Filling with concrete will prevent local buckling. General column buckling also needs to be checked and will usually be a maximum at a short distance below the mudline.

Installation may control the minimum wall thickness. The hammer blows develop high compressive waves which travel down the pile, reflecting from the tip in amplified compression when high tip resistance is encountered. When sustained hard driving with large hammers is anticipated, the minimum pile wall thickness should be $t = 6.35 + D/100$ where t and D are in millimeters. The drivability of a tubular pile is enhanced by increasing the wall thickness. This reduces the time of driving and enables greater penetration to be achieved.

During installation, the weight of the hammer and appurtenances may cause excessive bending if the pile is being installed on a batter. Hydraulic hammers usually are fully supported on the pile, whereas steam hammers and diesel hammers are partially supported by the crane.

If the pile is cleaned out during driving in order to enable the desired penetration to be achieved, external soil pressures may develop high circumferential compression stresses. These interact with the axial driving stresses and may lead to local buckling.

The tip of the pile is subject to very high stresses, especially if the pile encounters boulders or must be seated in rock. This may lead to distortion of the tip, which is then amplified during successive blows. In extreme cases, the tip may "tear" or may "accordion" in a series of short local axial buckles. Cast steel driving shoes may be employed in such cases; they are usually made of steels of high toughness as well as high yield strength. The pile head also must be thick enough to withstand both the local buckling and the bursting stresses due to Poisson's effect.

The transition between sections of different pile wall thickness must be carefully detailed. In general, the change in thickness should not be more than 12 mm at a splice and the thicker section should be beveled on a 1:4 slope.

3.2.4 Transportation and Upending of Piles

Tubular piles may be transported by barge. For loading, they are often simply rolled onto the barge, then blocked and chained down. They may also be transported by self-flotation. The ends are bulk-headed during deployment. The removal of the bulkheads can impose serious risks if not carefully planned. One end should be lifted above water for removal of that bulkhead, then the other. If one bulkhead is to be removed underwater by a diver, the water inside must first be equalized with the outside water; otherwise the rush of water will suck the diver into the pipe. Upending will produce high bending moments which limit the length of the sections of a long pile (Figure 3.4). Otherwise the pile may be buckled.

FIGURE 3.4 Large-diameter tubular steel pile being positioned.

3.2.5 Driving of Piles

The driving of large-diameter tubular piles [2] is usually done by a very large pile hammer. The required size can be determined by both experience and the use of a drivability analysis, which incorporates the soil parameters.

Frequently, the tubular pile for a bridge pier is too long or too heavy to install as a single section. Hence, piles must be spliced during driving. To assist in splicing, stabbing guides may be preattached to the tip of the upper segment, along with a backup plate. The tip of the upper segment should be prebeveled for welding.

Splicing is time-consuming. Fortunately, on a large-diameter pile of 2 to 4 m diameter, there is usually space to work two to three crews concurrently. Weld times of 4 to 8 h may be required. Then the pile must cool down (typically 2 h) and NDT performed. Following this, the hammer must be repositioned on top of the pile. Thus a total elapsed time may be 9 to 12 h, during which the skin friction on the pile sides "sets up," increasing the driving resistance and typically requiring a number of blows to break the pile loose and resume penetration.

When very high resistance is encountered, various methods may be employed to reduce the resistance so that the design pile tip may be reached. Care must be taken that these aids do not lessen the capacity of the pile to resist its design loads.

High resistance of the tubular pile is primarily due to plugging of the tip; the soil in the tip becomes compacted and the pile behaves as a displacement pile instead of cutting through the soil. The following steps may be employed.

FIGURE 3.5 Arrangement of internal jet piping and "spider" struts in large-diameter tubular pile.

1. *Jetting internally to break up the plug, but not below the tip.* The water level inside must be controlled, i.e., not allowed to build up much above the outside water level, in order to prevent piping underneath. Although a free jet or arrangement of jets may be employed, a very effective method is to manifold a series of jets around the circumference and weld the down-going pipes to the shell (Figure 3.5). Note that these pipes will pick up parasitic stresses under the pile hammer blows.

2. *Clean out by airlift.* This is common practice when using large-diameter tubular piles for bridge piers but has serious risks associated with it. The danger arises from the fact that an airlift can remove water very rapidly from the pile, creating an unbalanced head at the tip, and allowing run-in of soil. Such a run-in can result in major loss of resistance, not only under the tip in end bearing but also along the sides in skin friction.

   Unfortunately, this problem has occurred on a number of projects! The prevention is to have a pump operating to refill the pile at the same rate as the airlift empties it — a very difficult matter to control. If structural considerations allow, a hole can be cut in the pile wall so that the water always automatically balances. This, of course, will only be effective when the hole is below water. The stress concentrations around such a hole need to be carefully evaluated. Because of the risks and the service consequences of errors in field control, the use of an airlift is often prohibited. The alternative method, one that is much safer, is the use of a grab bucket (orange peel bucket) to remove the soil mechanically. Then, the water level can be controlled with relative ease.

3. *Drilling ahead a pilot hole, using slurry.* If the pile is kept full of slurry to the same level as the external water surface, then a pilot hole, not to exceed 75% of the diameter, may be drilled ahead from one to two diameters. Centralizers should be used to keep the drilled hole properly aligned. Either bentonite or a polymer synthetic slurry may be used. In soils such as stiff clay or where a binder prevents sloughing, seawater may be used. Reverse circulation is important to prevent erosion of the soils due to high-velocity flow. Drilling ahead is typically alternated with driving. The final seating should be by driving beyond the tip of the drilled hole to remobilize the plug resistance.

4. *External jetting.* External jetting relieves the skin friction during driving but sometimes permanently reduces both the lateral and axial capacity. Further, it is of only secondary benefit as compared with internal jetting to break up the plug. In special cases, it may still be employed. The only practicable method to use with long and large tubular piles is to weld the piping on the outside or inside with holes through the pile wall. Thus, the external jetting resembles that used on the much larger open caissons. As with them, low-pressure, high-volume water flow is most effective in reducing the skin friction. After penetration to the tip, grout may be injected to partially restore the lateral and axial capacity.

3.2.6 Utilization of Piles in Bridge Piers

There are several possible arrangement for tubular piles when used for bridge piers. These differ in some cases from those used in offshore platforms.

1. The pile may be driven to the required penetration and left with the natural soil inside. The upper portion may then be left with water fill or, in some cases, be purposely left empty in order to reduce mass and weight; in this case it must be sealed by a tremie concrete plug. To ensure full bond with the inside wall, that zone must be thoroughly cleaned by wire brush on a drill stem or by jet.

For piles fixed at their head, at least 2 diameters below the footing are filled with concrete to resist local buckling. Studs are installed in this zone to ensure shear transfer.

2. The pile, after driving to final penetration, is cleaned out to within one diameter of the tip. The inside walls are cleaned by wire brush or jet. A cage of reinforcing steel may be placed to augment the bending strength of the tubular shell. Centralizers should be used to ensure accurate positioning. The pile is then filled with tremie concrete. Alternatively, an insert steel tubular with plugged tip may be installed with centralizers, and the annular space filled with tremie grout. The insert tubular may need to be temporarily weighted and/or held down to prevent flotation in the grout.

Complete filling of a tubular pile with concrete is not always warranted. The heat of hydration is a potential problem, requiring special concrete mix design and perhaps precooling.

The reasons for carrying out this practice, so often adopted for bridge piers although seldom used in offshore structures, are

a. Concern over corrosion loss of the steel shell over the 100-year lifetime;
b. A need to ensure positively the ability of the permanent plug to sustain end bearing;
c. Prevention of local buckling near the mudline and at the pile head;
d. To obtain the benefits of composite behavior in stiffness and bending capacity.

If no internal supplemental reinforcement is required, then the benefits of (b), (c), and (d) may be achieved by simple filling with tremie concrete. To offset the heat of hydration, the core may be placed as precast concrete blocks, subsequently grouted into monolithic behavior. Alternatively, an insert pile may be full length. In this case, only the annulus is completely filled. The insert pile is left empty except at the head and tip.

The act of cleaning out the pile close to the tip inevitably causes stress relaxation in the soil plug below the clean-out. This will mean that under extreme axial compression, the pile will undergo a small settlement before it restores its full resistance. To prevent this, after the concrete plug has hardened, grout may be injected just beneath the plug, at a pressure that will restore the compactness of the soil but not so great as to pipe under the tip or fracture the foundation, or the pile may be re-seated by driving.

3. The tubular pile, after being installed to design penetration, may be filled with sand up to two diameters below the head, then with tremie concrete to the head. Reinforcing steel may be placed in the concrete to transfer part of the moment and tension into the footing block. Studs may be pre-installed on that zone of the pile to ensure full shear transfer. The soil and sand plug will act to limit local buckling at the mudline under extreme loads.

4. A socket may be drilled into rock or hard material beyond the tip of the driven pile, and then filled with concrete. Slurry is used to prevent degradation of the surface of the hole and sloughing. Seawater may be used in some rocks but may cause slaking in others such as shales and siltstone. Bentonite slurry coats the surface of the hole; the hole should be flushed with seawater just before concreting. Synthetic slurries are best, since they react in the presence of the calcium ion from the concrete to improve the bond. Synthetic polymer slurries biodegrade and thus may be environmentally acceptable for discharge into the water.

When a tubular pile is seated on rock and the socket is then drilled below the tip of the pile, it often is difficult to prevent run-in of sands from around the tip and to maintain proper circulation. Therefore, after landing, a hole may be drilled a short distance, for example, with a churn drill or down-the-hole drill, and then the pile reseated by the pile hammer.

Either insert tubulars or reinforcing steel cages are placed in the socket, extending well up into the pile. Tremie concrete is then placed to transfer the load in shear. In the case where a tubular insert pile is used, its tip may be plugged. Then grout may be injected into the annular space to transfer the shear.

Grout should not be used to fill sockets of large-diameter tubulars. The heat of hydration will damage the grout, reducing its strength. Tremie concrete should be used instead, employing small-size coarse aggregate, e.g., 15 mm, to ensure workability and flowability.

Although most sockets for offshore bridge piers have been cylindrical extensions of the tubular pile, in some offshore oil platforms belled footings have been constructed to transfer the load in end bearing. Hydraulically operated belling tools are attached to the drill string. Whenever transfer in end bearing is the primary mechanism, the bottom of the hole must be cleaned of silt just prior to the placement of concrete.

3.2.7 Prestressed Concrete Cylinder Piles

As an alternative to steel tubular piling, prestressed concrete cylinder piles have been used for a number of major overwater bridges, from the San Diego–Coronado and Dunbarton Bridges in California to bridges across Chesapeake Bay and the Yokohama cable-stayed bridge (Figures 3.6 and 3.7). Diameters have ranged from 1.5 to 6 m and more. They offer the advantage of durability and high axial compressive capacity. To counter several factors producing circumferential strains, especially thermal strains, spiral reinforcement of adequate cross-sectional area is required. This spiral reinforcement should be closely spaced in the 2-m zone just below the pile cap, where sharp reverse bending occurs under lateral loading.

Pile installation methods vary from driving and jetting of the smaller-diameter piles to drilling in the large-diameter piling (Figure 3.8).

FIGURE 3.6 Large-diameter prestressed concrete pile, Napa River Bridge, California.

FIGURE 3.7 Prestressed concrete cylinder pile for Oosterschelde Bridge, the Netherlands.

FIGURE 3.8 Installing concrete cylinder pile by internal excavation, jetting, and pull-down force from barge.

3.2.8 Footing Blocks

The footing block constructed at the top of large-diameter tubular piles serves the purpose of transmitting the forces from the pier shaft to the piles. Hence, it is subjected to large shears and significant moments. The shears require extensive vertical reinforcement, for both global shear (from the pier shaft) and local shear (punching shear from the piles). Large concentrations of reinforcement are required to distribute the moments. Post-tensioned tendons may be effectively utilized.

Although the primary forces typically produce compression in the upper surface of the footing block, secondary forces and particularly high temporary stresses caused by the heat of hydration produce tension in the top surface. Thus, adequate horizontal steel must be provided in the top and bottom in both directions.

The heat of hydration of the cemetitious materials in a large footing block develops over a period of several days. Due to the mass of the block, the heat in the core may not dissipate and return to ambient for several weeks.

The outside surface meantime has cooled and contracted, producing tension which often leads to cracking. Where inadequate reinforcement is provided, the steel may stretch beyond yield, so that the cracks become permanent. If proper amounts of reinforcement are provided, then the cracking that develops will be well distributed, individual cracks will remain small, and the elastic stress in the reinforcement will tend to close the cracks as the core cools.

Internal laminar cracking may also occur, so vertical reinforcement and middepth reinforcement should also be considered.

Footing blocks may be constructed in place, just above water, with precast concrete skirts extending down below low water in order to prevent small boats and debris from being trapped below. In this case, the top of the piles may be exposed at low water, requiring special attention to the prevention of corrosion.

Footing blocks may be constructed below water. Although cofferdams may be employed, the most efficient and economical way is usually to prefabricate the shell of the footing block. This is then floated into place. Corner piles are then inserted through the structure and driven to grade. The prefabricated box is then lowered down by ballasting, supported and guided by the corner piles. Then the remaining piles are threaded through holes in the box and driven. Final connections are made by tremie concrete.

Obviously, there are variations of the above procedure. In some cases, portions of the box have been kept permanently empty, utilizing their buoyancy to offset part of the deadweight.

Transfer of forces into the footing block requires careful detailing. It is usually quite difficult to transfer full moment by means of reinforcing inside the pile shell. If the pile head can be dewatered, reinforcing steel bars can be welded to the inside of the shell. Cages set in the concrete plug at the head may employ bundled bars with mechanical heads at their top. Alternatively the pile may be extended up through the footing block. Shear keys can be used to transfer shear. Post-tensioning tendons may run through and around the pile head.

3.3 Cofferdams for Bridge Piers

3.3.1 Description

The word *cofferdam* is a very broad term to describe a construction that enables an underwater site to be dewatered. As such, cofferdams can be large or small. Medium-sized cofferdams of horizontal dimensions from 10 to 50 m have been widely used to construct the foundations of bridge piers in water and soft sediments up to 20 m in depth; a few have been larger and deeper (Figure 3.9). Typical bridge pier cofferdams are constructed of steel sheet piles supported against the external pressures by internal bracing.

A few very large bridge piers, such as anchorages for suspension bridges, have utilized a ring of self-supporting sheet pile cells. The interior is then dewatered and excavated to the required depth. A recent such development has been the building of a circular ring wall of concrete constructed by the slurry trench method (Figures 3.10 and 3.11). Concrete cofferdams have also used a ring wall of precast concrete sheet piles or even cribs.

3.3.2 Design Requirements

Cofferdams must be designed to resist the external pressures of water and soil [3]. If, as is usual, a portion of the external pressures is designed to be resisted by the internal passive pressure of the soil, the depth of penetration must be selected conservatively, taking into account a potential sudden reduction in passive pressure due to water flow beneath the tip as a result of unbalanced water pressures or jetting of piles.

FIGURE 3.9 Large steel sheet pile cofferdam for Second Delaware Memorial Bridge, showing bracing frames.

FIGURE 3.10 Slurry wall cofferdam for Kawasaki Island ventilation shaft, Trans-Tokyo Bay tunnels and bridge.

The cofferdam structure itself must have adequate vertical support for self-load and equipment under all conditions.

In addition to the primary design loads, other loading conditions and scenarios include current and waves, debris and ice, overtopping by high tides, flood, or storm surge. While earthquake-induced loads, acting on the hydrodynamic mass, have generally been neglected in the past, they are now often being considered on major cofferdams, taking into account the lower input accelerations appropriate for the reduced time of exposure and, where appropriate, the reduced consequences.

Operating loads due to the mooring of barges and other floating equipment alongside need to be considered. The potential for scour must be evaluated, along with appropriate measures to reduce the scour. When the cofferdam is located on a sloping bank, the unbalanced soil loads need to be properly resisted. Accidental loads include impact from boats and barges, especially those working around the site.

The cofferdam as a whole must be adequately supported against the lateral forces of current waves, ice, and moored equipment, as well as unbalanced soil loads. While a large deep-water cofferdam appears to be a rugged structure, when fully excavated and prior to placement of the tremie concrete seal, it may

FIGURE 3.11   Concrete ring wall cofferdam constructed by slurry trench methods.

FIGURE 3.12   Dewatering the cofferdam for the main tower pier, Second Delaware Memorial Bridge.

be too weak to resist global lateral forces. Large tubular piles, acting as spuds in conjunction with the space-frame or batter piles may be needed to provide stability.

FIGURE 3.13 Pumped-out cofferdam showing tremie concrete seal and predriven steel H-piles.

The cofferdam design must be such as to integrate the piling and footing block properly. For example, sheet piles may prevent the installation of batter piles around the periphery. To achieve adequate penetration of the sheet piles and to accommodate the batter piles, the cofferdam may need to be enlarged. The arrangement of the bracing should facilitate any subsequent pile installation.

To enable dewatering of the cofferdam (Figure 3.12), a concrete seal is constructed, usually by the tremie method. This seal is designed to resist the hydrostatic pressure by its own buoyant weight and by uplift resistance provided by the piling, this latter being transferred to the concrete seal course by shear (Figure 3.13).

In shallow cofferdams, a filter layer of coarse sand and rock may permit pumping without a seal. However, in most cases, a concrete seal is required. In some recent construction, a reinforced concrete footing block is designed to be constructed underwater, to eliminate the need for a separate concrete seal. In a few cases, a drainage course of stone is placed below the concrete seal; it is then kept dewatered to reduce the uplift pressure. Emergency relief pipes through the seal course will prevent structural failure of the seal in case the dewatering system fails.

The underwater lateral pressure of the fresh concrete in the seal course and footing block must be resisted by external backfill against the sheet piles or by internal ties.

3.3.3 Internally Braced Cofferdams

These are the predominant type of cofferdams. They are usually rectangular in shape, to accommodate a regular pattern of cross-lot bracing.

The external wall is composed of steel sheet piles of appropriate section modulus to develop bending resistance. The loading is then distributed by horizontal wales to cross-lot struts. These struts should be laid out on a plan which will permit excavation between them, to facilitate the driving of piling and to eliminate, as far as practicable, penetration of bracing through the permanent structure.

Wales are continuous beams, loaded by the uniform bearing of sheet piles against them. They are also loaded axially in compression when they serve as a strut to resist the lateral loads acting on them end-wise. Wales in turn deliver their normal loads to the struts, developing concentrated local bearing loads superimposed upon the high bending moments, tending to produce local buckling. Stiffeners are generally required.

While stiffeners are readily installed on the upperside, they are difficult to install on the underside and difficult to inspect. Hence, these stiffeners should be pre-installed during fabrication of the members.

The wales are restrained from global buckling in the horizontal plane by the struts. In the vertical plane they are restrained by the friction of the sheet piles, which may need to be supplemented by direct fixation. Blocking of timber or steel shims is installed between the wales and sheet piles to fit the irregularities in sheet pile installation and to fill in the needed physical clearances.

Struts are horizontal columns, subject to high axial loading, as well as vertical loads from self-weight and any equipment that is supported by them. Their critical concern is stability against buckling. This is countered in the horizontal plane by intersecting struts but usually needs additional support in the vertical plane, either by piling or by trussing two or more levels of bracing.

The orthogonal horizontal bracing may be all at one elevation, in which case the intersections of the struts have to be accommodated, or they may be vertically offset, one level resting on top of the other. This last is normally easier since, otherwise, the intersections must be detailed to transmit the full loads across the joint. This is particularly difficult if struts are made of tubular pipe sections. If struts are made of wide-flanged or H-section members, then it will usually be found preferable to construct them with the weak axis in the vertical plane, facilitating the detailing of strut-to-strut intersections as well as strut-to-wale intersections. In any event, stiffeners are required to prevent buckling of the flanges.

For deep-water piers, the cofferdam bracing is best constructed as a space-frame, with two or more levels joined together by posts and diagonals in the vertical plane. This space-frame may be completely prefabricated and set as a unit, supported by vertical piles. These supporting piles are typically of large-diameter tubular members, driven through sleeves in the bracing frame and connected to it by blocking and welding.

The setting of such a space-frame requires a very large crane barge or equivalent, with both adequate hoisting capacity and reach. Sometimes, therefore, the bracing frame is made buoyant, to be partially or wholly self-floating. Tubular struts can be kept empty and supplemental buoyancy can be provided by pontoons.

Another way to construct the bracing frame is to erect one level at a time, supported by large tubular piles in sleeves. The lower level is first erected, then the posts and diagonal bracing in the vertical plane. The lower level is then lowered by hoists or jacks so that the second level can be constructed just above water and connections made in the dry.

A third way is to float in the prefabricated bracing frame on a barge, drive spud piles through sleeves at the four corners, and hang the bracing frame from the piles. Then the barge is floated out at low tide and the bracing frame lowered to position.

3.3.4 Circular Cofferdams

Circular cofferdams are also employed, with ring wales to resist the lateral forces in compression. The dimensions are large, and the ring compression is high. Unequal loading is frequently due to differential soil pressures. Bending moments are very critical, since they add to the compression on one side. Thus the ring bracing must have substantial strength against buckling in the horizontal plane.

3.3.5 Excavation

Excavation should be carried out in advance of setting the bracing frame or sheet piles, whenever practicable. Although due to side slopes the total volume of excavation will be substantially increased, the work can be carried out more efficiently and rapidly than excavation within a bracing system.

When open-cut excavation is not practicable, then it must be carried out by working through the bracing with a clamshell bucket. Struts should be spaced as widely as possible so as to permit use of a large bucket. Care must be taken to prevent impact with the bracing while the bucket is being lowered and from snagging the bracing from underneath while the bucket is being hoisted. These accidental loads may be largely prevented by temporarily standing up sheet piles against the bracing in the well being excavated, to act as guides for the bucket.

Except when the footing course will be constructed directly on a hard stratum or rock, overexcavation by 1 m or so will usually be found beneficial. Then the overexcavation can be backfilled to grade by crushed rock.

3.3.6   Driving of Piles in Cofferdams

Pilings can be driven before the bracing frame and sheet piles are set. They can be driven by underwater hammers or followers. To ensure proper location, the pile driver should be equipped with telescopic leads, or a template be set on the excavated river bottom or seafloor.

Piling may alternatively be driven after the cofferdam has been installed, using the bracing frame as a template. In this case, an underwater hammer presents problems of clearance due to its large size, especially for batter piles. Followers may be used, or, often, more efficiently, the piles may be lengthened by splicing to temporarily extend all the way to above water. They are then cut off to grade after the cofferdam has been dewatered. This procedure obviates the problems occasioned if a pile fails to develop proper bearing since underwater splices are not needed. It also eliminates cutoff waste. The long sections of piling cutoff after dewatering can be taken back to the fabrication yard and re-spliced for use on a subsequent pier.

All the above assumes driven steel piling, which is the prevalent type. However, on several recent projects, drilled shafts have been constructed after the cofferdam has been excavated. In the latter case, a casing must be provided, seated sufficiently deep into the bottom soil to prevent run-in or blowout.

Driven timber or concrete piles may also be employed, typically using a follower to drive them below water.

3.3.7   Tremie Concrete Seal

The tremie concrete seal course functions to resist the hydrostatic uplift forces to permit dewatering. As described earlier, it usually is locked to the foundation piling to anchor the slab. It may be reinforced in order to enable it to distribute the pile loads and to resist cracking due to heat of hydration.

Tremie concrete is a term derived from the French to designate concrete placed through a pipe. The term has subsequently evolved to incorporate both a concrete mix and a placement procedure. Underwater concreting has had both significant successes and significant failures. Yet the system is inherently reliable and concrete equal or better than concrete placed in the dry has been produced at depths up to 250 m. The failures have led to large cost overruns due to required corrective action. They have largely been due to inadvertently allowing the concrete to flow through or be mixed with the water, which has caused washout of the cement and segregation of the aggregate.

Partial washout of cement leads to the formation of a surface layer of laitance which is a weak paste. This may harden after a period of time into a brittle chalklike substance.

The tremie concrete mix must have an adequate quantity of cementitious materials. These can be a mixture of portland cement with either fly ash or blast furnace slag (BFS). These are typically proportioned so as to reduce the heat of hydration and to promote cohesiveness. A total content of cementitious materials of 400 kg/m^3 (~700 lb/cy) is appropriate for most cases.

Aggregates are preferably rounded gravel so they flow more readily. However, crushed coarse aggregates may be used if an adequate content of sand is provided. The gradation of the combined aggregates should be heavy toward the sand portion — a 45% sand content appears optimum for proper flow. The maximum size of coarse aggregate should be kept small enough to flow smoothly through the tremie pipe and any restrictions such as those caused by reinforcement. Use of 20 mm maximum size of coarse aggregate appears optimum for most bridge piers.

A conventional water-reducing agent should be employed to keep the water/cementitious material ratio below 0.45. Superplasticizers should not normally be employed for the typical cofferdam, since the workability and flowability may be lost prematurely due to the heat generated in the mass concrete. Retarders are essential to prolong the workable life of the fresh mix if superplasticizers are used.

Other admixtures are often employed. Air entrainment improves flowability at shallow water depths but the beneficial effects are reduced at greater depths due to the increased external pressure. Weight to reduce uplift is also lost.

Microsilica may be included in amounts up to 6% of the cement to increase the cohesiveness of the mix, thus minimizing segregation. It also reduces bleed. Antiwashout admixtures (AWA) are also employed to minimize washout of cementitious materials and segregation. They tend to promote self-leveling and flowability. Both microsilica and AWA may require the use of superplasticizers in which case retarders are essential. However, a combination of silica fume and AWA should be avoided as it typically is too sticky and does not flow well.

Heat of hydration is a significant problem with the concrete seal course, as well as with the footing block, due to the mass of concrete. Therefore, the concrete mix is often precooled, e.g., by chilling of the water or the use of ice. Liquid nitrogen is sometimes employed to reduce the temperature of the concrete mix to as low as 5°C. Heat of hydration may be reduced by incorporating substantial amounts of fly ash to replace an equal portion of cement. BFS–cement can also be used to reduce heat, provided the BFS is not ground too fine, i.e., not finer than 2500 $cm^{2/}g$ and the proportion of slag is at least 70% of the total.

The tremie concrete mix may be delivered to the placement pipe by any of several means. Pumping and conveyor belts are best because of their relatively continuous flow. The pipe for pumping should be precooled and insulated or shielded from the sun; conveyor belts should be shielded. Another means of delivery is by bucket. This should be air-operated to feed the concrete gradually to the hopper at the upper end of the tremie pipe. *Placement down the tremie pipe should be by gravity feed only* (Figure 3.14).

Although many placements of tremie concrete have been carried out by pumping, there have been serious problems in large placements such as cofferdam seals. The reasons include:

1. Segregation in the long down-leading pipe, partly due to formation of a partial vacuum and partly due to the high velocity;
2. The high pressures at discharge;
3. The surges of pumping.

FIGURE 3.14 Placing underwater concrete through hopper and tremie pipe, Verrazano Narrows Bridge, New York.

Since the discharge is into fresh concrete, these phenomena lead to turbulence and promote intermixing with water at the surface, forming excessive laitance.

These discharge effects can be contrasted with the smooth flow from a gravity-fed pipe in which the height of the concrete inside the tremie pipe automatically adjusts to match the external pressure of water vs. the previously placed concrete. For piers at considerable depths, this balance point will be about half-way down. The pipe should have an adequate diameter in relation to the maximum size of coarse aggregate to permit remixing: a ratio of 8 to 1 is the minimum. A slight inclination of the tremie pipe from the vertical will slow the feed of new concrete and facilitate the escape of entrapped air.

For starting the tremie concrete placement, the pipe must first be filled slightly above middepth. This is most easily done by plugging the end and placing the empty pipe on the bottom. The empty pipe must be negatively buoyant. It also must be able to withstand the external hydrostatic pressure as well as the internal pressure of the underwater concrete. Joints in the tremie pipe should be gasketed and bolted to prevent water being sucked into the mix by venturi action. To commence placement, with the tremie pipe slightly more than half full, it is raised 150 mm off the bottom. The temporary plug then comes off and the concrete flows out. The above procedure can be used both for starting and for resuming a placement, as, for example, when the tremie is relocated, or after a seal has been inadvertently lost.

The tremie pipe should be kept embedded in the fresh concrete mix a sufficient distance to provide backpressure on the flow (typically 1 m minimum), but not so deep as to become stuck in the concrete due to its initial set. This requires adjustment of the retarding admixture to match the rate of concrete placement and the area of the cofferdam against the time of set, keeping in mind the acceleration of set due to heat as the concrete hydrates.

Another means for initial start of a tremie concrete placement is to use a pig which is forced down the pipe by the weight of the concrete, expelling the water below. This pig should be round or cylindrical, preferably the latter, equipped with wipers to prevent leakage of grout and jamming by a piece of aggregate. An inflated ball, such as an athletic ball (volleyball or basketball) must never be used; these collapse at about 8 m water depth! A pig should not be used to restart a placement, since it would force a column of water into the fresh concrete previously placed.

Mixes of the tremie concrete described will flow outward on a slope of about 1 on 8 to 1 on 10. With AWAs, an even flatter surface can be obtained.

A trial batch with underwater placement in a shallow pit or tank should always be done before the actual placement of the concrete seal. This is to verify the cohesiveness and flowability of the mix. Laboratory tests are often inadequate and misleading, so a large-scale test is important. A trial batch of 2 to 3 m^3 has often been used.

The tremie concrete placement will exert outward pressure on the sheet piles, causing them to deflect. This may in turn allow new grout to run down past the already set concrete, increasing the external pressure. To offset this, the cofferdam can be partially backfilled before starting the tremie concreting and tied across the top. Alternatively, dowels can be welded on the sheets to tie into the concrete as it sets; the sheet piles then have to be left in place.

Due to the heat of hydration, the concrete seal will expand. Maximum temperature may not be achieved for several days. Cooling of the mass is gradual, starting from the outside, and ambient temperature may not be achieved for several weeks. Thus an external shell is cooling and shrinking while the interior is still hot. This can produce severe cracking, which, if not constrained, will create permanent fractures in the seal or footing. Therefore, in the best practice, reinforcing steel is placed in the seal to both provide a restraint against cracking and to help pull the cracks closed as the mass cools.

After a relatively few days, the concrete seal will usually have developed sufficient strength to permit dewatering. Once exposed to the air, especially in winter, the surface concrete will cool too fast and may crack. Placing insulation blankets will keep the temperature more uniform. They will, of course, have to be temporarily moved to permit the subsequent work to be performed.

3.3.8   Pier Footing Block

The pier footing block is next constructed. Reinforcement on all faces is required, not only for structural response but also to counteract thermal strains.

The concrete expands as it is placed in the footing block due to the heat of hydration. At this stage it is either still fresh or, if set, has a very low modulus. Then it hardens and bonds to the tremie concrete below. The lock between the two concrete masses is made even more rigid if piling protrudes through the top of the tremie seal, which is common practice. Now the footing block cools and tries to shrink but is restrained by the previously placed concrete seal. Vertical cracks typically form. Only if there is sufficient bottom reinforcement in both directions can this shrinkage and cracking be adequately controlled. Note that these tensile stresses are permanently locked into the bottom of the footing block and the cracks will not close with time, although creep will be advantageous in reducing the residual stresses.

After the footing block has hardened, blocking may be placed between it and the sheet piles. This, in turn, may permit removal of the lower level of bracing. As an alternative to bracing, the footing block may be extended all the way to the sheet piles, using a sheet of plywood to prevent adhesion.

3.3.9   Pier Shaft

The pier shaft is then constructed. Block-outs may be required to allow the bracing to pass through. The internal bracing is removed in stages, taking care to ensure that this does not result in overloading a brace above. Each stage of removal should be evaluated.

Backfill is then placed outside the cofferdam to bring it up to the original seabed. The sheet piles can then be removed. The first sheets are typically difficult to break loose and may require driving or jacking in addition to vibration. Keeping in mind the advantage of steel sheet piles in preventing undermining of the pier due to scour, as well as the fact that removal of the sheets always loosens the surrounding soil, hence reducing the passive lateral resistance, it is often desirable to leave the sheet piles in place below the top of the footing. They may be cut off underwater by divers; then the tops are pulled by vibratory hammers.

Antiscour stone protection is now placed, with an adequate filter course or fabric sheet in the case of fine sediments.

3.4   Open Caissons

3.4.1   Description

Open caissons have been employed for some of the largest and deepest bridge piers [4]. These are an extension of the "wells" which have been used for some 2000 years in India. The caisson may be constructed above its final site, supported on a temporary sand island, and then sunk by dredging out within the open wells of the caisson, the deadweight acting to force the caisson down through the overlying soils (Figure 3.15). Alternatively, especially in sites overlain by deep water, the caisson may be prefabricated in a construction basin, floated to the site by self-buoyancy, augmented as necessary by temporary floats or lifts, and then progressively lowered into the soils while building up the top.

Open caissons are effective but costly, due to the large quantity of material required and the labor for working at the overwater site. Historically, they have been the only means of penetrating through deep overlying soils onto a hard stratum or bedrock. However, their greatest problem is maintaining stability during the early phases of sinking, when they are neither afloat nor firmly embedded and supported. Long and narrow rectangular caissons are especially susceptible to tipping, whereas square and circular caissons of substantial dimensions relative to the water depth are inherently more stable. Once the caisson tips, it tends to drift off position. It is very difficult to bring it back to the vertical without overcorrecting.

FIGURE 3.15 Open-caisson positioned within steel jackets on "pens."

When the caisson finally reaches its founding elevation, the surface of rock or hard stratum is cleaned and a thick tremie concrete base is placed. Then the top of the caisson is completed by casting a large capping block on which to build the pier shaft.

3.4.2 Installation

The sinking of the cofferdam through the soil is resisted by skin friction along the outside and by bearing on the cutting edges. Approximate values of resistance may be obtained by multiplying the friction factor of sand on concrete or steel by the at-rest lateral force at that particular stage, $f = ØK_0wh^2$ where f is the unit frictional resistance, $Ø$ the coefficient of friction, w the underwater unit weight of sand, K_0 the at-rest coefficient of lateral pressure, and h the depth of sand at that level. f is then summed up over the embedded depth. In clay, the cohesive shear controls the "skin friction." The bearing value of the cutting edges is generally the "shallow bearing value," i.e., five times the shear strength at that elevation.

These resistances must be overcome by deadweight of the caisson structure, reduced by the buoyancy acting on the submerged portions. This deadweight may be augmented by jacking forces on ground anchors.

The skin friction is usually reduced by lubricating jets causing upward flow of water along the sides. Compressed air may be alternated with water through the jets; bentonite slurry may be used to provide additional lubrication. The bearing on the cutting edges may be reduced by cutting jets built into the walls of the caisson or by free jets operating through holes formed in the walls. Finally, vibration of the soils near and around the caisson may help to reduce the frictional resistance.

When a prefabricated caisson is floated to the site, it must be moored and held in position while it is sunk to and into the seafloor. The moorings must resist current and wave forces and must assist in maintaining the caisson stable and in a vertical attitude. This latter is complicated by the need to build up the caisson walls progressively to give adequate freeboard, which, of course, raises the center of gravity.

Current force can be approximately determined by the formula

$$F = CA\rho \frac{V^2}{2g}$$

where C varies from 0.8 for smooth circular caissons to 1.3 for rectangular caissons, A is the area, ρ is the density of water, and V is the average current over the depth of flotation. Steel sheet piles develop high drag, raising the value of C by 20 to 30%.

As with all prismatic floating structures, stability requires that a positive metacentric height be maintained. The formula for metacentric height, $\overline{GM}$, is

$$\overline{GM} = \overline{KB} - \overline{KG} + \overline{BM}$$

where $\overline{KB}$ is the distance from the base to the center of buoyancy, $\overline{KG}$ is the distance to the center of gravity, and $\overline{BM} = I/V$.

I is the moment of inertia on the narrowest (most sensitive) axis, while V is the displaced volume of water. For typical caissons, a $\overline{GM}$ of +1 m or more should be maintained.

The forces from mooring lines and the friction forces from any dolphins affect the actual attitude that the structure assumes, often tending to tip it from vertical. When using mooring lines, the lines should be led through fairleads attached near the center of rotation of the structure. However, this location is constantly changing, so the fairlead attachment points may have to be shifted upward from time to time.

Dolphins and "pens" are used on many river caissons, since navigation considerations often preclude mooring lines. These are clusters of piles or small jackets with pin piles and are fitted with vertical rubbing strips on which the caisson slides.

Once the caisson has been properly moored on location, it is ballasted down. As it nears the existing river or harbor bottom, the current flow underneath increases dramatically. When the bottom consists of soft sediments, these may rapidly scour away in the current. To prevent this, a mattress should be first installed.

Fascine mattresses of willow, bamboo, or wood with filter fabric attached are ballasted down with rock. Alternatively, a layer of graded sand and gravel, similar to the combined mix for concrete aggregate, can be placed. The sand on top will scour away, but the final result will be a reverse filter.

In order to float a prefabricated caisson to the site initially, false bottoms are fitted over the bottom of the dredging wells. These false bottoms are today made of steel, although timber was used on many of the famous open caissons from the 19th and the early part of the 20th centuries. They are designed to resist the hydrostatic pressure plus the additional force of the soils during the early phases of penetration. Once the caisson is embedded sufficiently to ensure stability, the false bottoms are progressively removed so that excavation can be carried out through the open wells. This removal is a very critical and dangerous stage, hazardous both to the caisson and to personnel. The water level inside at this stage should be slightly higher than that outside. Even then, when the false bottom under a particular well is loosened, the soil may suddenly surge up, trapping a diver. The caisson, experiencing a sudden release of bearing under one well, may plunge or tip.

Despite many innovative schemes for remote removal of false bottoms, accidents have occurred. Today's caissons employ a method for gradually reducing the pressure underneath and excavating some of the soil before the false bottom is released and removed. For such constructions, the false bottom is of heavily braced steel, with a tube through it, typically extending to the water surface. The tube is kept full of water and capped, with a relief valve in the cap. After the caisson has penetrated under its own weight and come to a stop, the relief valve is opened, reducing the pressure to the hydrostatic head only. Then the cap is removed. This is done for several (typically, four) wells in a balanced pattern. Then jets and airlifts may be operated through the tube to remove the soil under those wells. When the caisson has penetrated sufficiently far for safety against tipping, the wells are filled with water; the false bottoms are removed and dredging can be commenced.

3.4.3 Penetration of Soils

The penetration is primarily accomplished by the net deadweight, that is, the total weight of concrete steel and ballast less the buoyancy. Excavation within the wells is carried down in a balanced pattern until the bearing stratum is reached. Then tremie concrete is placed, of sufficient depth to transfer the design bearing pressures to the walls.

The term *cutting edge* is applied to the tips of the caisson walls. The external cutting edges are shaped as a wedge while the interior ones may be either double-wedge or square. In the past, concern over concentrated local bearing forces led to the practice of making the cutting edges of heavy and expensive fabricated steel. Today, high-strength reinforced concrete is employed, although if obstructions such as boulders, cobbles, or buried logs are anticipated or if the caisson must penetrate rock, steel armor should be attached to prevent local spalling.

The upper part of the caisson may be replaced by a temporary cofferdam, allowing the pier shaft dimensions to be reduced through the water column. This reduces the effective driving force on the caisson but maintains and increases its inherent stability.

The penetration requires the progressive failure of the soil in bearing under the cutting edges and in shear along the sides. Frictional shear on the inside walls is reduced by dredging while that on the outside walls is reduced by lubrication, using jets as previously described.

Controlling the penetration is an essentially delicate balancing of these forces, attempting to obtain a slight preponderance of sinking force. Too great an excess may result in plunging of the caisson and tipping or sliding sidewise out of position. That is why pumping down the water within the caisson, thus reducing buoyancy, is dangerous; it often leads to sudden inflow of water and soil under one edge, with potentially catastrophic consequences.

Lubricating jets may be operated in groups to limit the total volume of water required at any one time to a practicable pump capacity. In addition to water, bentonite may be injected through the lubricating jets, reducing the skin friction. Compressed air may be alternated with water jetting.

Other methods of aiding sinking are employed. Vibration may be useful in sinking the caisson through sands, especially when it is accompanied by jetting. This vibration may be imparted by intense vibration of a steel pile located inside the caisson or even by driving on it with an impact hammer to liquefy the sands locally.

Ground anchors inserted through preformed holes in the caisson walls may be jacked against the caisson to increase the downward force. They have the advantage that the actual penetration may be readily controlled, both regarding force exerted and displacement.

Since all the parameters of resistance and of driving force vary as the caisson penetrates the soil, and because the imbalance is very sensitive to relatively minor changes in these parameters, it is essential to plan the sinking process in closely spaced stages, typically each 2 to 3 m. Values can be precalculated for each such stage, using the values of the soil parameters, the changes in contact areas between soil and structure, the weights of concrete and steel and the displaced volume. These need not be exact calculations; the soil parameters are estimates only since they are being constantly modified by the jetting. However, they are valuable guides to engineering control of the operations.

There are many warnings from the writings of engineers in the past, often based on near-failures or actual catastrophes.

1. Verify structural strength during the stages of floating and initial penetration, with consideration for potentially high resistance under one corner or edge.
2. In removing false bottoms, be sure the excess pressure underneath has first been relieved.
3. Do not excavate below cutting edges.
4. Check outside soundings continually for evidence of scour and take corrective steps promptly.
5. Blasting underneath the cutting edges may blow out the caisson walls. Blasting may also cause liquefaction of the soils leading to loss of frictional resistance and sudden plunging. If blasting is needed, do it before starting penetration or, at least, well before the cutting edge reaches the hard strata so that a deep cushion of soil remains over the charges.
6. If the caisson tips, avoid drastic corrections. Instead, plan the correction to ensure a gradual return to vertical and to prevent the possibility of tipping over more seriously on the other side. Thus steps such as digging deeper on the high side and overballasting on the high side

FIGURE 3.16 Open caisson for Sunshine Bridge across the Mississippi River. This caisson tipped during removal of false bottoms and is shown being righted by jacking against dolphins.

should be last resorts and, then, some means mobilized to arrest the rotation as the caisson nears vertical. Jacking against an external dolphin is a safer and more efficient method for correcting the tipping (Figure 3.16).

Sinking the caisson should be a continuous process, since once it stops the soil friction and/or shear increase significantly and it may be difficult to restart the caisson's descent.

3.4.4 Founding on Rock

Caissons founded on bare rock present special difficulties. The rock may be leveled by drilling, blasting, and excavation, although the blasting introduces the probability of fractures in the underlying rock. Mechanical excavation may therefore be specified for the last meter or two. Rotary drills and underwater road-headers can be used but the process is long and costly. In some cases, hydraulic rock breakers can be employed; in other cases a hammer grab or star chisel may be used. For the Prince Edward Island Bridge, the soft rock was excavated and leveled by a very heavy clamshell bucket. Hydraulic backhoes and dipper dredges have been used elsewhere. A powerful cutter-head dredge has been planned for use at the Strait of Gibraltar.

3.4.5 Contingencies

The planning should include methods for dealing with contingencies. The resistance of the soil and especially of hard strata may be greater than anticipated. Obstructions include sunken logs and even sunken buried barges and small vessels, as well as cobbles and boulders. The founding rock or stratum may be irregular, requiring special means of excavating underneath the cutting edge at high spots or filling in with concrete in the low spots. One contingency that should always be addressed is what steps to take if the caisson unexpectedly tips.

Several innovative solutions have been used to construct caissons at sites with especially soft sediments. One is a double-walled self-floating concept, without the need for false bottoms. Double-walled caissons of steel were used for the Mackinac Strait Bridge in Michigan. Ballast is progressively filled into the double-wall space while dredging is carried out in the open wells.

In the case of extremely soft bottom sediments, the bottom may be initially stabilized by ground improvement, for example, with surcharge by dumped sand, or by stone columns, so that the caisson may initially land on and penetrate stable soil. Great care must, of course, be exercised to maintain control when the cutting edge breaks through to the native soils below, preventing erratic plunging.

This same principle holds true for construction on sand islands, where the cutting edge and initial lifts of the caisson may be constructed on a stratum of gravel or other stable material, then the caisson sunk through to softer strata below. Guides or ground anchors will be of benefit in controlling the sinking operation.

3.5 Pneumatic Caissons

3.5.1 Description

These caissons differ from the open caisson in that excavation is carried out beneath the base in a chamber under air pressure. The air pressure is sufficient to offset some portion of the ambient hydrostatic head at that depth, thus restricting the inflow of water and soil.

Access through the deck for workers and equipment and for the removal of the excavated soil is through an airlock. Personnel working under air pressure have to follow rigid regimes regarding duration and must undergo decompression upon exit. The maximum pressures and time of exposure under which personnel can work is limited by regulations. Many of the piers for the historic bridges in the United States, e.g., the Brooklyn Bridge, were constructed by this method.

3.5.2 Robotic Excavation

To overcome the problems associated with working under air pressure, the health hazards of "caisson disease," and the high costs involved, robotic cutters [5] have been developed to excavate and remove the soil within the chamber without human intervention. These were recently implemented on the piers of the Rainbow Suspension Bridge in Tokyo (Figures 3.17 and 3.18).

The advantage of the pneumatic caisson is that it makes it possible to excavate beneath the cutting edges, which is of special value if obstructions are encountered. The great risk is of a "blowout" in which the air escapes under one edge, causing a rapid reduction of pressure, followed by an inflow of water and soil, endangering personnel and leading to sudden tilting of the caisson. Thus, the use of pneumatic caissons is limited to very special circumstances.

FIGURE 3.17 Excavating within pressurized working chamber of pneumatic caisson for Rainbow Suspension Bridge, Tokyo. (Photo courtesy of Shiraishi Corporation.)

FIGURE 3.18 Excavating beneath cutting edge of pneumatic caisson. (Photo courtesy of Shiraishi Corporation.)

3.6 Box Caissons

3.6.1 Description

One of the most important developments of recent years has been the use of box caissons, either floated in or set in place by heavy-lift crane barges [4]. These box caissons, ranging in size from a few hundred tons to many thousands of tons, enable prefabrication at a shore site, followed by transport and installation during favorable weather "windows" and with minimum requirements for overwater labor. The development of these has been largely responsible for the rapid completion of many long overwater bridges, cutting the overall time by a factor of as much as three and thus making many of these large projects economically viable.

The box caisson is essentially a structural shell that is placed on a prepared underwater foundation. It is then filled with concrete, placed by the tremie method previously described for cofferdam seals. Alternatively, sand fill or just ballast water may be used.

Although many box caissons are prismatic in shape, i.e., a large rectangular base supporting a smaller rectangular column, others are complex shells such as cones and bells. When the box caisson is seated on a firm foundation, it may be underlain by a meter or two of stone bed, consisting of densified crushed rock or gravel that has been leveled by screeding. After the box has been set, underbase grout is often injected to ensure uniform bearing.

3.6.2 Construction

The box caisson shell is usually the principal structural element although it may be supplemented by reinforcing steel cages embedded in tremie concrete. This latter system is often employed when joining a prefabricated pier shaft on top of a previously set box caisson.

Box caissons may be prefabricated of steel; these were extensively used on the Honshu–Shikoku Bridges in Japan (Figures 3.19 and 3.20). After setting, they were filled with underwater concrete, in earlier cases using grout-intruded aggregate, but in more recent cases tremie concrete.

For reasons of economy and durability, most box caissons are made of reinforced concrete. Although they are therefore heavier, the concurrent development of very large capacity crane barges and equipment has made their use fully practicable. The weight is advantageous in providing stability in high currents and waves.

FIGURE 3.19 Steel box caisson being positioned for Akashi Strait Bridge, Japan, despite strong currents.

FIGURE 3.20 Akashi Strait Suspension Bridge is founded on steel box caissons filled with tremie concrete.

3.6.3 Prefabrication Concepts

Prefabrication of the box caissons may be carried out in a number of interesting ways. The caissons may be constructed on the deck of a large submersible barge. In the case of the two concrete caissons for the Tsing Ma Bridge in Hong Kong, the barge then moved to a site where it could submerge to launch the caissons. They were floated to the site and ballasted down onto the predredged rock base. After sealing the perimeter of the cutting edge, they were filled with tremie concrete.

In the case of the 66 piers for the Great Belt Western Bridge in Denmark, the box caissons were prefabricated on shore in an assembly-line process (Figures 3.21 and 3.22). They were progressively moved out onto a pier from which they could be lifted off and carried by a very large crane barge to their site. They were then set onto the prepared base. Finally, they were filled with sand and antiscour stone was placed around their base.

FIGURE 3.21 Prefabrication of box caisson piers for Great Belt Western Bridge, Denmark.

FIGURE 3.22 Prefabricated concrete box caissons are moved by jacks onto pier for load-out.

FIGURE 3.23 Large concrete box caissons fabricated in construction basin for subsequent deployment to site by self-flotation, Great Belt Eastern Bridge, Denmark.

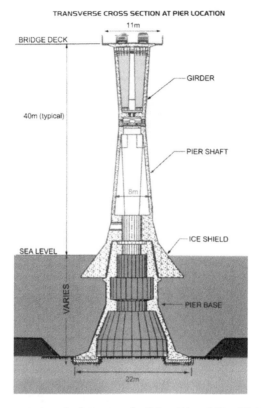

FIGURE 3.24 Schematic representation of substructure for Prince Edward Island Bridge, Canada. Note ice shield, designed to reduce forces from floating ice in Northumberland Strait.

A similar procedure has been followed for the approach piers on the Oresund Bridge between Sweden and Denmark and on the Second Severn Bridge in Southwest England. For the Great Belt Eastern Bridge, many of the concrete box caissons were prefabricated in a construction basin (Figure 3.23). Others were fabricated on a quay wall.

For the Prince Edward Island Bridge, bell-shaped piers, with open bottom, weighing up to 8000 tons, were similarly prefabricated on land and transported to the load-out pier and onto a barge, using transporters running on pile-supported concrete beams (Figures 3.24 through 3.26). Meanwhile, a shallow trench had been excavated in the rock seafloor, in order to receive the lower end of the bell. The bell-shaped shell was then lowered into place by the large crane barge. Tremie concrete was placed to fill the peripheral gap between bell and rock.

3.6.4 Installation of Box Caissons by Flotation

Large concrete box caissons have been floated into location, moored, and ballasted down onto the prepared base (Figure 3.27). During this submergence, they are, of course, subject to current, wave, and wind forces. The moorings must be sufficient to control the location; "taut moorings" are therefore used for close positioning.

The taut moorings should be led through fairleads on the sides of the caisson, in order to permit lateral adjustment of position without causing tilt. In some cases where navigation requirements prevent the use of taut moorings, dolphins may be used instead. These can be faced with a vertical rubbing strip or master pile. Tolerances must be provided in order to prevent binding.

Stability is of critical importance for box caissons which are configured such that the water plane diminishes as they are submerged. It is necessary to calculate the metacentric height, $\overline{GM}$, at every change in horizontal cross section as it crosses the water plane, just as previously described for open caissons.

FIGURE 3.25 Prefabrication of pier bases, Prince Edward Island Bridge, Canada.

FIGURE 3.26 Prefabricated pier shaft and icebreaker, Prince Edward Island Bridge, Canada.

During landing, as during the similar operation with open caissons, the current under the caisson increases, and scour must be considered. Fortunately, in the case of box caissons, they are being landed either directly on a leveled hard stratum or on a prepared bed of densified stone, for which scour is less likely.

As the base of the caisson approaches contact, the prism of water trapped underneath has to escape. This will typically occur in a random direction. The reaction thrust of the massive water jet will push the caisson to one side. This phenomenon can be minimized by lowering the last meter slowly.

FIGURE 3.27 Prefabricated box caisson is floated into position, Great Belt Eastern Bridge, Denmark. Note temporary cofferdam above concrete caisson.

Corrections for the two phenomena of current scour and water-jet thrust are in opposition to one another, since lowering slowly increases the duration of exposure to scour. Thus it is essential to size and compact the stone of the stone bed properly and also to pick a time of low current, e.g., slack tide for installation.

3.6.5   Installing Box Caissons by Direct Lift

In recent years, very heavy lift equipment has become available. Jack-up, floating crane barges, and catamaran barges, have all been utilized (Figures 3.28 through 3.31). Lifts up to 8000 tons have been made by crane barge on the Great Belt and Prince Edward Island Bridges.

The box caissons are then set on the prepared bed. Where it is impracticable to screed a stone bed accurately, landing seats may be preset to exact grade under water and the caisson landed on them, and tremie concrete filled in underneath.

Heavy segments, such as box caissons, are little affected by current — hence can be accurately set to near-exact location in plan. Tolerances of the order of 20 to 30 mm are attainable.

FIGURE 3.28 Installing precast concrete box caissons for Second Severn Crossing, Bristol, England. Extreme tidal range of 10 m and high tidal current imposed severe demands on installation procedures and equipment.

FIGURE 3.29 Lifting box caisson from quay wall on which it was prefabricated and transporting it to site while suspended from crane barge.

FIGURE 3.30 Setting prefabricated box caisson on which is mounted a temporary cofferdam, Great Belt Eastern Bridge, Denmark.

FIGURE 3.31 Setting 7000-ton prefabricated box caisson, Great Belt Western Bridge, Denmark.

3.6.6 Positioning

Electronic distance finders (EDF), theodolites, lasers, and GPS are among the devices utilized to control the location and grade. Seabed and stone bed surveys may be by narrow-beam high-frequency sonar and side-scan sonar. At greater depths, the sonar devices may be incorporated in an ROV to get the best definition.

3.6.7 Grouting

Grouting or concreting underneath is commonly employed to ensure full bearing. It is desirable to use low-strength, low-modulus grout to avoid hard spots. The edges of the caisson have to be sealed by penetrating skirts or by flexible curtains which can be lowered after the caisson is set in place, since otherwise the tremie concrete will escape, especially if there is a current. Heavy canvas or submerged nylon, weighted with anchor chain and tucked into folds, can be secured to the caisson during prefabrication. When the caisson is finally seated, the curtains can be cut loose; they will restrain concrete or grout at low flow pressures. Backfill of stone around the edges can also be used to retain the concrete or grout.

Heat of hydration is also of concern, so the mix should not contain excessive cement. The offshore industry has developed a number of low-heat, low modulus, thixotropic mixes suitable for this use. Some of them employ seawater, along with cement, fly ash, and foaming agents. BFS cement has also been employed.

Box caissons may be constituted of two or more large segments, set one on top of the other and joined by overlapping reinforcement encased in tremie concrete. The segments often are match-cast to ensure perfect fit.

3.7 Present and Future Trends

3.7.1 Present Practice

There is a strong incentive today to use large prefabricated units, either steel or concrete, that can be rapidly installed with large equipment, involving minimal on-site labor to complete. On-site operations, where required, should be simple and suitable for continuous operation. Filling prefabricated shells with tremie concrete is one such example.

Two of the concepts previously described satisfy these current needs. The first, a box caisson — a large prefabricated concrete or steel section — can be floated in or lifted into position on a hard seafloor. The second, large-diameter steel tubular piles, can be driven through soft and variable soils found in a competent stratum, either rock or dense soils. These tubular piles are especially suitable for areas of high seismicity, where their flexibility and ductility can be exploited to reduce the acceleration transmitted to the superstructure (Figure 3.32). In very deep water, steel-framed jackets may be employed to support the piles through the water column (Figure 3.33). The box caisson, conversely, is most suitable to resist the impact forces from ship collision. The expanding use of these two concepts is leading to further incremental improvements and adaptations which will increase their efficiency and economy.

Meanwhile cofferdams and open caissons will continue to play an important but diminishing role. Conventional steel sheet pile cofferdams are well suited to shallow water with weak sediments, but involve substantial overwater construction operations.

3.7.2 Deep Water Concepts

The Japanese have had a study group investigating concepts for bridge piers in very deep water, and soft soils. One initial concept that has been pursued is that of the circular cofferdam constructed of concrete by the slurry wall process [3]. This was employed on the Kobe anchorage for the Akashi

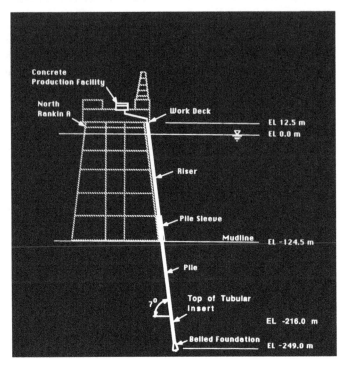

FIGURE 3.32 Conceptual design for deep-water bridge pier, utilizing prefabricated steel jacket and steel tubular pin piles.

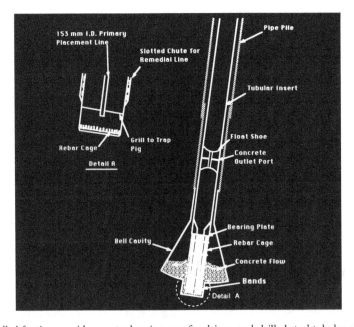

FIGURE 3.33 Belled footing provides greater bearing area for driven-and-drilled steel tubular pile.

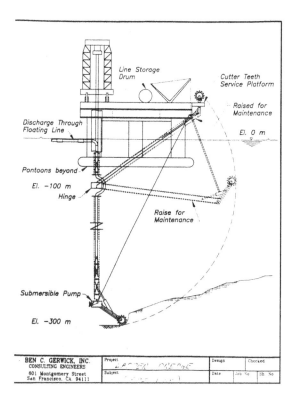

FIGURE 3.34 Concept for preparation of seabed for seating of prefabricated box piers in 300 m water depth, Strait of Gibraltar.

Strait Bridge and on the Kawasaki Island ventilation structure for the Trans-Tokyo Bay tunnels, the latter with a pumped-out head of 80 m, in extremely soft soils in a zone of high seismicity (see Section 3.3).

Floating piers have been proposed for very deep water, some employing semisubmersible and tension "leg-platform" concepts from the offshore industry. While technically feasible, the entire range of potential adverse loadings, including accidental flooding, ship impact, and long-period swells, need to be thoroughly considered. Tethered pontoons of prestressed concrete have been successfully used to support a low-level bridge across a fjord in Norway.

Most spectacular of all proposed bridge piers are those designed in preliminary feasibility studies for the crossing of the Strait of Gibraltar. Water depths range from 305 m for a western crossing to 470 m for a shorter eastern crossing. Seafloor soils are highly irregular and consist of relatively weak sandstone locally known as flysch. Currents are strong and variable. Wave and swell exposure is significant. For these depths, only offshore platform technology seems appropriate.

Both steel jackets with pin piles and concrete offshore structures were investigated. Among the other criteria that proved extremely demanding were potential collision by large crude oil tankers and, below water, by nuclear submarines.

These studies concluded that the concrete offshore platform concept was a reasonable and practicable extension of current offshore platform technology. Leveling and preparing a suitable foundation is the greatest challenge and requires the integration and extension of present systems of dredging well beyond the current state of the art (Figure 3.34). Conceptual systems for these structures have been developed which indicate that the planned piers are feasible by employing an extension of the concepts successfully employed for the offshore concrete platforms in the North Sea (Figure 3.35).

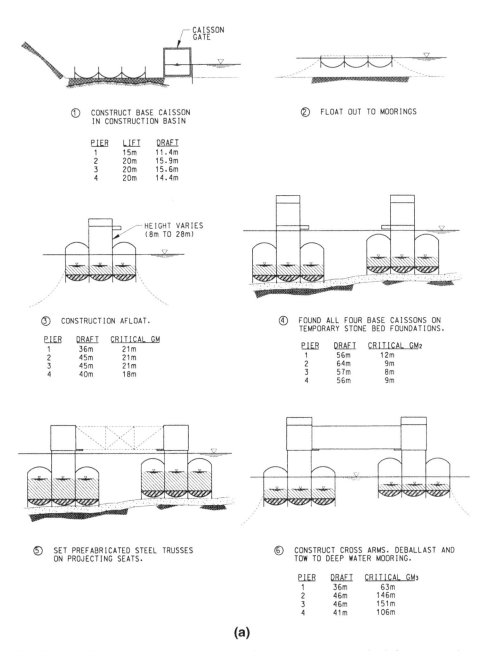

(a)

FIGURE 3.35 (a,b,c) Fabrication and installation concept for piers in 300 m water depth for crossing of Strait of Gibraltar.

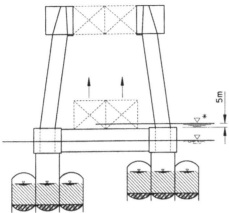

⑦ BALLAST DOWN TO MID-HEIGHT OF CROSS-ARMS.
CONSTRUCT ALL 4 SHAFTS AND CONICAL TOPS.
RAISE PLATFORM FABRICATION TRUSSES AND
SECURE TO SHAFT TOP CONES AND UPPER
FALSEWORK (TRUSS WEIGTHS '- 2000T)

PIER	DRAFT*	CRITICAL GM_4
1	96m	22m
2	106m	34m
3	95m	36m
4	97m	27m

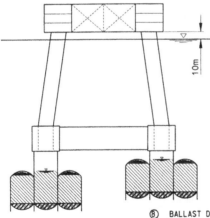

⑧ BALLAST DOWN TO DEEP DRAFT. COMPLETE
UPPER PLATFORM, VERTICAL CROSS WALLS
FORMED WITH REBAR TIED BUT NO CONCRETE
PLACED.

PIER	DRAFT*	CRITICAL GM_5
1	103m	5m
2	281m	21m
3	246m	16m
4	196m	6m

(b)

FIGURE 3.35 (continued)

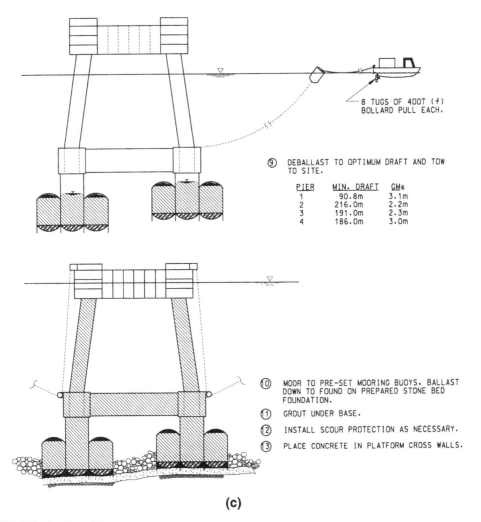

(c)

FIGURE 3.35 (continued)

References

1. API-RP2A, Recommended Practice for Planning, Designing and Constructing Fixed Offshore Platforms, 1993.
2. Gerwick, Ben C., Jr., *Construction of Offshore Structures*, John Wiley & Sons, New York, 1986.
3. Ratay, R. T., Ed., Cofferdams, in *Handbook of Temporary Structures in Construction*, 2nd ed., McGraw-Hill, New York, 1996, chap. 7.
4. O'Brien, J. J., Havers, J. A., and Stubbs, F. W., *Standard Handbook of Heavy Construction*, 3rd ed., McGraw-Hill, New York, 1996, chap. B-11 (Marine Equipment), chapter D-4 (Cofferdams and Caissons).
5. Shiraishi, S., Unmanned excavation systems in pneumatic caissons, in *Developments in Geotechnical Engineering*, A.A. Balkema, Rotterdam, 1994.

4

Bridge Construction Inspection

Mahmoud Fustok
*California Department
of Transportation*

Masoud Alemi
*California Department
of Transportation*

4.1 Introduction

Bridge construction inspection provides quality assurance for building bridges and plays a very important role in the bridge industry. Bridge construction involves two types of structures: permanent and temporary structures. Permanent structures, including foundations, abutments, piers, columns, wingwalls, superstructures, and approach slabs, are those that perform the structural functions of a bridge during its service life. Temporary structures, such as shoring systems, guying systems, forms and falsework, are those that support the permanent structure during its erection and construction.

This chapter discusses inspection principles, followed by the guidelines for inspecting materials, construction operations, component construction, and temporary structures. It also touches on safety considerations and documentation.

4.2 Inspection Objectives and Responsibilities

4.2.1 Objectives

The objective of construction inspection is to ensure that the work is being performed according to the project plans, specifications [1,2], and the appropriate codes including AASHTO and AWS [3,4] as necessitated by the project. The specifications describe the expected quality of materials; standard methods of work; methods and frequency of testing; and the variation or tolerance allowed. Design and construction of temporary structures should also meet the requirements of construction manuals [5–8].

4.2.2 Responsibilities of the Inspector

The inspector's primary responsibility is to make sure that permanent structures are constructed in accordance with project plans and specifications and to ensure that the operations and/or products meet the quality standard. The inspector is also responsible to determine the design adequacy of temporary structures proposed for use by the contractor. The qualified inspector should have a thorough knowledge of specifications and should exercise good judgment. The inspector should keep a detailed diary of daily observations, noting particularly all warnings and instructions given to the contractor. The inspector should maintain continual communication with the contractor and resolve issues before they become problems.

4.3 Material Inspection

4.3.1 Concrete

At the beginning of the project, a set of concrete mix designs should be proposed by the contractor for use in the project. These mix designs are based on the specification requirements, the desired workability of the mix, and availability of local resources. The proposed concrete mix designs should be reviewed and approved by the inspector.

Method and frequency of sampling and testing of concrete are covered in the specifications. Concrete cylinders are sampled, cured, and tested to determine their compressive strength. The following tests are conducted to check some other concrete properties including:

- Cement content;
- Cleanness value of the coarse aggregate;
- Sand equivalent of the fine aggregate;
- Fine, coarse, and combined aggregate grading;
- Uniformity of concrete.

4.3.2 Reinforcement

Reinforcing steel properties and fabrication should conform to the specifications. A Certificate of Compliance and a copy of the mill test report for each heat and size of reinforcing steel should be furnished to the inspector. These reports should show the physical and chemical analysis of the reinforcing bars.

Bars should not be bent in a manner that damages the material. Check for cracking on Grade 60 reinforcing steel where radii of hooks have been bent too tight. Bars with kinks or improper bends must not be used in the project. Hooks and bends must conform to the specifications.

For epoxy-coated reinforcing bars, a Certificate of Compliance conforming to the specifications must be furnished for each shipment. Epoxy coating material must be tested for specification compliance. Bars are tested for coating thickness and for adhesion requirements. Any damage to the coating caused by shipment and/or installation must be repaired with patching material compatible with the coating material.

4.3.3 Structural Steel

The contractor furnishes to the inspector a copy of all mill orders, certified mill test reports, and a Certificate of Compliance for all fabricated structural steel to be used in the work [9]. In addition, the test reports for steels with specified impact values include the results of Charpy V-notch impact tests.

Structural steel which is used to fabricate fracture critical members should meet the more stringent Charpy V-notch requirements [4].

4.4 Operation Inspection

4.4.1 Layout and Grades

Bridge inspectors spend a major part of their time making sure that the structure is being built at the correct locations and elevations as shown on the project plans. Lines and grades are provided at reference points by surveyors. Based on these reference points the contractor establishes the lines and grades for building the structure. Horizontal alignment usually consists of a series of circular curves connected with tangents. Vertical alignment usually consists of a series of parabolic curves that also are connected with tangents. When a bridge is on a horizontal curve, its cross section is sloped to counteract the centrifugal forces. This slope is called the superelevation. Therefore, the inspector uses the reference points and basic geometric principles for horizontal and vertical curves to check the bridge geometry. The following layout and grades are to be checked in the field:

1. Pile locations and cutoff elevations;
2. Footings location and grades;
3. Column pour grades;
4. Abutments and wingwalls pour grades;
5. Falsework grade points;
6. Lost deck grade points;
7. Overhang jacks and edge of deck.

The deck contour sheet (DCS) is a scaled topographic plan that shows the top of the bridge deck elevations [10]. With some manipulation, the DCS provides elevations for various components and construction stages of a bridge, including abutments, columns, falsework, lost deck, and edge of deck. Keeping tight control on alignment and grades with the DCS will produce smooth vehicle ridability and an aesthetically pleasing bridge.

Falsework (FW) grades are adjusted to meet the soffit elevations. FW points should be placed at locations where the FW will be adjusted, and may be shot at each stage. For example, FW points should not be set on girder centerlines of a cast-in-place prestressed box-girder bridge; otherwise they will be covered by the prestress ducts.

Lost deck dowels (LDD) are the control points to construct the top deck of the bridge. Accurate field layout and DCS layout of the LDD points are essential. Before the soffit and stem pour, the contractor places LDD at predetermined points. After the girder pour, the inspector measures the elevations of the LDD and compares them to those picked from the DCS; then the amount of adjustment needed to the deck grades is calculated.

Inspection of the edge of deck (EOD) grades is very critical for controlling and smooth operation of deck-finishing tools. Additionally, EOD controls the thickness of the concrete slab. It is recommended to locate EOD grade points at points of form adjustment.

FW bents are erected at elevations higher than the theoretical grades to offset the anticipated FW settlements. For FW used to construct cast-in-place concrete bridges, camber strips are usually placed on top of FW stringers to offset (1) deflection of the FW beam under its own weight and the actual load imposed, (2) difference between beam profile and bridge profile grade, and (3) difference between beam profile and any permanent camber.

Steel plate girders are fabricated with built-in upward camber to offset vertical deflections due to the girder weight and other dead loads such as deck and barrier, deflection caused by concrete creep, and to provide any vertical curve required by the profile. The total values are usually called the web camber. Header camber is the web camber minus the girder deflection due to its own weight. If an FW bent is needed to erect a field splice for a steel girder, the grades of the FW bent should be set to no-load elevations. No-load elevation is equal to the plan elevation plus the sum of deflections minus the depth of the superstructure.

4.4.2   Concrete Pour

Ready-mixed concrete is usually delivered to the field by concrete trucks. Each load of ready-mixed concrete should be accompanied by a ticket showing volume of concrete, mix number, time of batch, and reading of the revolution counter. The inspector should ensure the concrete meets the specification requirements regarding:

- Elapse time of batch
- Number of drum revolutions
- Concrete temperature
- Concrete slump (penetration)

Addition of water to the delivered concrete, at the job site, should be approved by the inspector and it should follow the specifications.

The inspector should ensure that concrete is consolidated using vibrators by methods that will not cause segregation of aggregates and will result in a dense homogeneous concrete free of voids and rock pockets. The vibrator should not be dragged horizontally over the top of the concrete surface. Special care must be taken in vibrating areas where there is a high concentration of reinforcing steel. To prevent concrete from segregation caused by excessive free fall, a double-belting hopper or an elephant trunk should be used to guide concrete placing for piles, foundations, and walls.

While placing concrete into a footing, pouring concrete at one location and using vibrators to spread it should not be allowed since it causes aggregate segregation. Care should be taken when topping off column pour to the proper grade. For a cast-in-place box-girder bridge, columns are usually poured 30 mm higher than the theoretical grade to help butt soffit plywood and hide the joint.

While placing concrete, soffit thickness and height of concrete in stems should be checked. Concrete in bent caps should be placed at the proper grade to allow room for minimum concrete clearance over the main cap top reinforcement as shown on the project plans. The appropriate rebar concrete cover is essential in preventing steel rusting.

During bridge deck placement, the pour front should not exceed 3 to 4 m ahead of the finishing machine. Application of curing compound should be performed by power-operated equipment, and it should follow the finishing machine closely.

4.4.3   Reinforcement Placing

Reinforcement should be properly placed as shown on the plans in terms of grade, size, quantity, and location of steel rebar. Reinforcement should be firmly and securely held in position by wiring at intersections and splices with wire ties. The "pigtails" on wire ties should be flattened so that they maintain the minimum concrete coverage from formed surfaces. Rust bleeding can occur from pigtails which extend to the concrete surface.

Positioning of reinforcing steel in the forms is usually accomplished by the use of precast mortar blocks. The use of metal, plastic, and wooden support chairs is normally not permitted. For soffit and deck, blocks should be sufficient to keep mats off plywood a distance equal to the specified rebar clearance. For short- to medium-height columns, reinforcement clearance can be checked from the top of the form with a mirror or flashlight. For foundations, adequate blocking should be provided to hold the bottom mat in proper position. If column steel rests on the bottom mat, extra blocking may be required. Reinforcing steel should be protected from bond breaker substances.

Splicing of reinforcing bars can be done by lapping, butt welding, mechanical butt splice, or mechanical lap splicing.

For lap splicing A615 steel rebar, the following splicing length is recommended [1]:

For Grade 300:

- Bar No. 25 and smaller is 30D
- Bar Nos. 30 and 35 is 45D

For Grade 400:

- Bar No. 25 or smaller is 45D
- Bar Nos. 30 and 35 is 60D

where D is diameter of the smaller bar joined.

Reinforcing bars larger than No. 35 should not be spliced by lapping.

For welded rebar, welds should be the right size and be free of cracks, lack of fusion, undercutting, and porosity. Butt welds should be made with multiple passes while flare welds may be made in one pass. For quality assurance, radiographic examinations might be performed on full-penetration butt-welded splices. Radiographs can be made by either X ray or gamma ray.

For mechanical butt splices, splicing should be performed in accordance with the manufacturer's recommendations. Tests of sample splices should be done for quality assurance.

Reinforcement for abutment and wingwalls should be checked before forms are buttoned up.

Proper reinforcing steel placement in deck, especially truss bars, is very important, since the moment-carrying capacity of a bridge deck is greatly sensitive to the effective depth of the section. String lines are used between grade points to check steel clearance and deck thickness.

4.4.4 Welding of Structural Steel

The inspector is responsible for the quality assurance inspection (QAI) of all welding. The inspector should ascertain that equipment, procedures, and techniques are in accordance with specifications. Contract specifications usually refer to welding codes such as AWS D1.5 [4]. All welding should be performed in accordance with an approved welding procedure and by a certified welder. All welding materials such as electrodes, fluxes, and shielding gases must be properly packaged, stored, and dried. Quality of the welds is largely dependent on the welding equipment. The equipment should be checked to ensure that it is in good working condition. Travel speed and rate of flow of shielding gases should be monitored. The actual heat input should not exceed the maximum heat input that was tested and approved. Preheat and interpass temperature specifications should be adhered to as they affect cooling rate and heat input. When postheat is specified, the temperature and duration must be monitored.

Base metal at the welding area (root face, groove face) must be within allowable roughness tolerances. All mill scale should be removed from surfaces where girder web to flange welds are made.

Inspectors must verify that all nondestructive testing (NDT) has been performed and passed the specified requirements [11,12]. The inspector should maintain a record of all locations of inspected areas with the NDT report and findings, together with the method of repairs and NDT test results of weld repairs.

Fillet weld profile should be within final dimensional requirements for leg and throat size and surface contour. Bumps and craters due to starts and stops, weld rollover, and insufficient leg and throat must be ground and repaired to acceptable finish.

4.4.5 High-Strength Bolts

The contact surfaces of all high-strength bolted connections should be thoroughly cleaned of rust, mill scale, dirt, grease, paint, lacquer, or other material. Before the installation of fasteners, the inspector should check the marking, surface condition of bolts, nuts, and washers for compliance with the specifications. Nuts and bolts that are not galvanized should be clean and dry or lightly lubricated. Nuts for high-strength galvanized bolts should be overtapped after galvanizing, and then treated with a lubricant [13].

High-strength bolts may be tensioned by use of a calibrated power wrench, a manual torque wrench, the turn-of-nut method, or by tightening and using a direct tension indicator. The inspector should observe calibration and testing procedures required to confirm that the selected procedure is properly applied and the required tensions are provided [14]. The inspector should monitor the installation in the work to ensure that the selected procedure, as demonstrated in the testing, is routinely properly applied.

To inspect completed joints, the following procedure is recommended [15]. A representative sample of bolts from the diameter, length, and grade of the bolts used in the work should be tightened in the tension-measuring device by any convenient means to an initial condition equal to approximately 15% of the required fastener tension and then to the minimum tension specified. Tightening beyond the initial condition must not produce greater nut rotation than 1.5 times that permitted in the specifications. The inspection wrench should be applied to the tightened bolts and the torque necessary to turn the nut or head 5° should be determined.

The Coronet load indicator (CLI) is a simple and accurate aid for tightening and inspecting high-strength bolts. The CLI is a hardened round washer with bumps on one face. As the bolt and nut are tightened, the clamping force flattens the bumps, placed against the underside of the bolt head. The nut should be tightened until the gap is reduced to 0.38 mm. This requirement applies to both A325 and A490 bolts. Once the gap is reduced to the required dimension, the bolt and nut are properly tightened. Visual gap inspection is usually adequate by comparing them against gaps which were checked with the feeler gauge. All CLI for A325 are round. CLI for A490 has three ears or the letter V stamped at three places.

Reuse of ASTM A490 and galvanized ASTM A490 bolts is not allowed. Reuse of ASTM A325 bolts may be allowed if it is approved by the inspector. Reuse does not include retightening bolts which may have been loosened by the tightening of adjacent bolts.

4.5 Component Inspection

4.5.1 Foundation

Conventional bridge foundations can be classified in three types: (1) spread footing foundation, (2) pile foundation, (3) special-case foundations such as pile shafts, tiebacks, soil nails, and tie-downs. The first two types of foundations are more common, and therefore, their inspections will be discussed. Problems that may be encountered during foundation construction should be discussed with the designer and the engineering geologist who performed the foundation studies.

Spread Footing Foundations

Spread footing foundations support the load by bearing directly on the foundation stratum. The conformity of the foundation material with the log of the test boring should be checked. Additionally, the bearing surface should be free of disturbed material and be compacted if it is necessary.

Foundations are poured using forms (Figure 4.1) or using the soil as a form in neat-line excavation method. The "neat-line" excavation method is usually done for column and retaining wall footings. The toe of the retaining wall should be placed against undisturbed material. Depth and dimensions of footing need to be checked. Standing water and all sloughed material in the excavation must be removed prior to placement of concrete into the foundation. The foundation material should be wet down but not saturated. Footings more than 760 mm vertical dimension, and with a top layer of reinforcement, should be reconsolidated by a vibrator for a depth of 300 mm. Reconsolidation should be done not less than 15 min after the initial leveling of the top of the foundation has been completed. A curing compound will usually be used on top of the footing. If the construction joint is sandblasted before the completion of the curing period, the exposed area should be cured using an alternative method for the remaining time.

Pile Foundations

Pile foundations transmit design loads into adjacent soil through pile friction, end bearing, or both. Tops of piles are never exact. Determine the pile with highest elevation and block up bottom-mat steel reinforcement to horizontal accordingly. The highest pile may have to be cut off if the grade is unreasonable.

FIGURE 4.1 Foundation forms.

There are two major types of piles: cast-in-drilled-hole piles and driven piles.

Cast-in-Drilled-Hole Concrete Piles
For cast-in-drilled-hole piles, the inspector should check the following:

- Diameter, depth, and straightness of drilled holes;
- Cleanness of the bottom of holes from water and loose materials.

Material encountered during drilling should be compared with that shown on the log of test borings. If there is a significant difference, the designer should be informed.

For 400-mm-diameter piles, the top 4500 mm of concrete should be vibrated; for larger-diameter piles the full length should be vibrated [16].

Using steel casing is one method to prevent soil cave-ins and intrusion of groundwater. Casing is pulled when placing concrete keeping its bottom below the concrete surface. Waiting too long to pull the casing may cause the concrete to set up and may lead to the following problems:

- The concrete comes up with the casing.
- The casing cannot be removed.
- The concrete may not fill the voids left by the casing.

Use of a concrete mix with fluidity at the high end of the allowable range will help to mitigate these problems.

A slurry displacement method can be used to prevent cave-in of unstable soil and intrusion of groundwater into the drilled hole. The drilling slurry remains in the drilled hole until it is displaced by concrete. Concrete is placed using a delivery system with rigid tremie tube or a rigid pump tube, starting at the bottom of the drilled hole. Sampling and testing of drilling slurry is an important quality control requirement. The following properties of drilling slurries should be monitored: density, sand content, pH value, and viscosity [16].

Inspection tubes are installed inside the spiral or hoop reinforcement in a straight alignment in order to facilitate pile testing. Inspection tubes permit the insertion of a testing probe that measures the density of the pile concrete. A radiographic technique, commonly called gamma ray scattering, is used to measure the density. If the pile is accepted, the inspection tubes are cleaned and filled with grout.

Driven Piles

Prior to start-up of a pile-driving operation, the inspector should check the hammer type and the pile size. Piles should be marked for logging. During the pile-driving operation, the inspector should monitor plumbness or batter of the pile, and log the pile penetration.

Charts of calibration curves are developed for different pile capacity and hammer. By using the energy theory, a penetration-per-blow chart which corresponds to the specified capacity of the piles can be developed. One of the commonly used formulas to determine the bearing capacity of a pile is the ENR formula:

$$p = \frac{E}{6(s + 2.54)} \ q \qquad (4.1)$$

where

  p = safe load in kilonewtons
  E = manufacturer's rating for energy developed by the hammer in joules
  s = average penetration per blow in millimeters

On a large structure which necessitates long piles to be driven in order to satisfy the design load-bearing requirements, monitoring tests may be performed to determine the pile-driving chart.

For pipe steel piles, piles should have the specified diameter, length, and wall thickness as shown on the project plans. If the piles are to be spliced, welding should be performed by a certified welder and the quality of the welded joints should be checked. The method of pipe splicing should be in accordance with project plans and specification requirements. Steel shells may be driven open or closed ended. For shells that are driven open ended, soil should be augured out, and the pile should be cleaned before it is filled with concrete.

For precast concrete piles, the following should be checked:

• Pile is free of damage or cracks;
• Size and length of piles;
• Age of pile, minimum 14 days before it is allowed to be driven [1].

If piles are driven through new embankment greater than 1500 mm, predrilling is required [1]. Problems with the driving operation can be categorized into three types:

1. Hard driving occurs if the soil is too dense or the hammer does not have enough energy to drive the pile. This problem can be solved by predrilling, jetting, or using a larger low-velocity hammer. If a pile is undergoing hard driving and suddenly experiences a large movement, this could indicate a fracture of the pile belowground. In this case the pile should be extracted and replaced or a replacement pile should be driven next to it.
2. Soft pile occurs when the pile is driven to the specified tip elevation but has not attained the specified bearing capacity. This pile is set for a minimum of 12 h, then retapped. If the retapped pile will not attain the bearing capacity, then the contractor has to furnish longer piles.
3. Alignment of piles: if the pile begins to move out of plumbness, correction should be made. The pile may have to be pulled and redriven.

4.5.2 Concrete Columns and Pile Shaft

For medium to long columns, column cages should be held at the top with a crane until footing concrete is set as insurance for the guying system (Figure 4.2). For extremely tall columns, a crane holds the cage until the column is poured.

For fixed columns, make sure ties to the bottom mat are placed in accordance with the project plans to prevent the cage moving during placement of concrete into the footing. For pinned columns, check key details and verify that adequate blocking is provided to support the steel cage at the proper height above a key.

FIGURE 4.2 Crane used to hold the column cage; column is being cured with plastic wrap.

Check the sequence of attaching and removing the guying system to ensure it is done in accordance with the approved plans. Location of utilities should be checked prior to forming a column.

Column forms are usually removed a few days after the concrete pour. Columns should be covered with plastic wrap until they are cured (Figure 4.2).

Pile shafts are encountered in bedrock material usually close to canyons or hillside areas with limited room for footing foundations. They are primarily a cast-in-drilled-hole footing with neat-line excavation. The column has the same size extension as the pile shaft or a slightly smaller section. Shoring is required in areas that are not solid rock and any excavation out of the neat area should be filled with concrete. Since blasting is the most common method of excavation, extreme caution is necessary to protect workers and the public.

4.5.3 Abutment and Wingwalls

Abutment and wingwalls are normally formed and poured at the same time for seat-type abutments (Figure 4.3). For the diaphragm abutments of cast-in-place, prestressed-concrete box girders, wingwalls should be placed after stressing. Utility openings in wingwalls and/or abutments should be checked in accordance with the project plans. Bearing pad and internal key layout are usually checked after pour strips are in place.

4.5.4 Superstructure

The following is a summary of the construction inspection for concrete box and steel plate girders which are commonly used for building short- to medium-size bridges.

Cast-in-Place Concrete Box Girder

Concrete box girders are constructed in two stages. First, FW is erected and stem and soffit are formed and poured (Figure 4.4). Second, lost deck is built and then the concrete for the bridge deck is poured (Figure 4.5). For prestressed concrete, once the bridge deck is cured, the frame is prestressed, and then the FW is removed. Erection of stem and soffit will be discussed here, and bridge prestressing and deck erection will be discussed in following sections.

FIGURE 4.3 Wingwall and abutment construction.

FIGURE 4.4 Bridge stem and soffit forms for concrete box girder.

The following items need to be checked during construction of soffit and stems:

- Size of camber strips placed on top of FW stringers;
- Location of utility and/or soffit access openings and the corresponding reinforcing details;
- Location and elevation of block-outs for deck drain pipes;
- Size and profile of the prestressing ducts;
- Smoothness of duct-to-flare connection at bearing plate for alignment of line profile (Figure 4.6);
- Tattletales readings; readings exceeding those anticipated should be investigated to determine if they are due to FW failure or due to excessive soil settlement; corrective measure should be taken accordingly.

FIGURE 4.5 Bridge deck reinforcement with bidwell finishing machine.

FIGURE 4.6 Bridge soffit and girder reinforcements with prestressing duct profile.

Inspectors should make sure that

- Trumpets are properly secured to the bearing plates;
- Ducts and intermediate vents are secured in place;
- Proper gap is maintained between snap ties and ducts to prevent any damage to prestressing ducts;
- Tendon openings are sealed to prevent water or debris from entering;
- Size and location of prestress bearing plates at abutment or hinge diaphragms are in accordance with the approved prestressing working drawings;

FIGURE 4.7 Prestressing strands and anchor set.

- Elastomeric bearing pads are installed properly at abutments and the remainder of the abutment seat area is covered with expanded polystyrene of the same thickness as the pads and that joints are sealed with tape to prevent grout leakage;
- Curing compound is applied appropriately and on time.

Due to the tremendous amount of force involved in prestressing a concrete bridge, careful consideration should be given to this operation. Caution should be exercised around the prestressing jack during the stressing operation and around ducts while they are not grouted (Figure 4.7).

Prior to placement of tendons in prestressing ducts, the following should be verified:

- Ducts are unobstructed and free of water and debris;
- Strands are free of rust.

Prior to stressing, the contractor should submit the required calibration curves for specific jack/gauge combinations. It is the inspector's responsibility to determine

- That tendons are installed in accordance with plans;
- Whether stressing should be done from one end or two ends;
- That stressing sequence is performed in accordance with plans.

During stressing, strands are painted on both ends to check slippage. Prestressing calibration curves are plotted, and measured elongation is compared to calculated elongation using the actual area and modulus of elasticity [17].

During the grouting of tendons, the following needs to be checked:

- Water/cement ratio;
- Efflux time;
- That grout is continuously agitated.

Steel Girder

Structural steel is usually inspected at the fabrication site. In the field, girders should be checked for damage that may have occurred during transportation to the job. All bearing assemblies should be set

FIGURE 4.8   Bearing assembly of steel plate girder.

level and to the elevations shown on the plans [18]. Full bearing on concrete should be obtained under bearing assemblies (Figure 4.8).

Fracture-critical members (FCM) are tension members or tension components of bending members, the failure of which may result in collapse of the bridge. FMCs are usually identified on the plans or described in the contract documents. FCMs are subject to the additional provisions of Section 12 of AASHTO/AWS D1.5. All welds to FCMs are considered fracture critical and should conform to the requirements of the fracture control plan.

All surfaces of structural steel that are to be painted should be blast-cleaned to produce a dense, uniform surface with an angular anchor pattern [19]. On the same day that blast cleaning is done, structures should be painted with undercoats prior to their erection. Steel surfaces that are inaccessible for painting after erection should be fully painted before they are erected. Steel areas where paint has been damaged due to erection should be cleaned and painted with undercoats before the application of any subsequent paint. Subsequent painting should not be performed until the cleaned surfaces are dry. Succeeding applications of paint should be of such shade to provide contrast with the paint being covered. Paint thickness is measured with a thickness gauge that is calibrated by a magnetic film.

Precast Prestressed Girders

Like steel beams, precast prestressed girders are inspected at the fabricator site. Precast concrete members are usually cured by the water or steam method. Upon arrival to the project site, girders should be inspected for damage that may have occurred while being transported to the project site.

Precast girders should be handled, transported, and erected using extreme care to avoid twisting, racking, or other distortion that would result in cracking or damage. Girders are placed on elastomeric pads at certain locations shown in the plans. Girders are braced and held together by temporary wooden blocking.

The top of the girder elevation is determined by profiling each girder. The profile grade will determine the location of the finished grade of the top slab and the location of the slab forms. After girders have been profiled and finished grades have been determined, placing forms for the top slab can start. Prestressed concrete panels are a type of slab form that is left in place and becomes the bottom part of the concrete slab. They are normally 100 mm thick [20], rectangular shaped, and vary in width and length.

Concrete Deck

The bridge top deck is probably the most critical part in terms of smooth vehicle ridability and aesthetically pleasing bridge. Smoothness of the bridge surface and the approach slabs should be tested by a bridge profilograph. Two profiles will be obtained in each lane. Surfaces that fail to conform to the smoothness tolerances should be ground until these tolerances are achieved. Grinding should not reduce the concrete cover on reinforcing steel to less than 40 mm [1]. The following items need to be checked during the construction of bridge deck:

1. Adequacy of sandblasting on top of girders of cast-in-place concrete bridges;
2. Tops of girders are free of dust;
3. Block-outs for joint seal assemblies;
4. Overhang chamfer and screed pipes have smooth line;
5. Clearance of finishing machine roller to the steel mat (see Figure 4.5), and height of deck drain inlets in relation to finishing machine roller;
6. Tattletales for additional FW settlement;
7. The curing rugs periodically for dampness after completion of deck pour.

Inspectors should ensure that:

1. Soffit vents, access openings, drains, and their support systems are clear from steel rebar and located in accordance with the project plans [21];
2. Application of rugs or mats is begun within 4 h after completion of deck finishing, and no later than the following morning (Figure 4.9);
3. Prior to prestressing or release of FW, deck surface is inspected for crack intensity to ensure that it meets the allowable tolerance set forth in the specification.

4.6 Temporary Structures

4.6.1 Falsework

FW is used to provide temporary support to the superstructure during construction. To construct a cast-in-place concrete bridge, a complete FW system is needed. Such a system consists of FW foundation,

FIGURE 4.9 Bridge deck is being cured with moist rugs.

FIGURE 4.10 FW system for cast-in-place concrete bridge.

bents, stringers, joists, and forms (Figure 4.10). To erect a steel girder, a simple FW system might be needed. Such a system consists of FW bent and foundation.

FW is designed by the contractor and approved by the inspector. FW is designed to resist vertical loads, as well as longitudinal and transverse horizontal loads. The inspector needs to have a basic understanding of design of timber members, steel members, cables, and to be familiar with application of the FW manual to check the submitted calculations and plans accordingly.

Temporary bracing or other means are needed to hold FW in stable condition during erection and removal.

Typical FW includes timber pads, corbels, steel cells, timber or steel posts, steel or timber FW bent caps, steel stringers, and timber joists. Cables are usually used to resist lateral load in the longitudinal direction. Timber, cable, or steel bar bracing is used to resist lateral loads in the transverse direction. Forms are placed on top of the joist.

When FW pad foundations are used, soil-bearing capacity may be approximated based on observed soil classification or by performing soil load tests. FW pads must be placed on a level and firm material. Pad foundations should be protected from flooding and from undermining by surface runoff. Continuous pads should be inspected to ensure the pad joints are located according to the approved FW plans.

Timber FW materials should be inspected for defects. Bolts and/or nails can be used to connect timber framing. Inspectors should check edge distance, end distance, and minimum spacing as required by the FW manual [5]. For nailed or spiked connections, ensure that adequate penetrations are provided. If the grades of FW bent are adjusted at the bottom of the post, check diagonal bracing and its connection for any distortion caused by differential movement. Wedging might be needed to ensure full bearing at contact surfaces.

The use of worn or kinked cable should not be permitted. Cables should be looped around a thimble with a minimum diameter corresponding to the cable diameter. Proper clip installation is critical and should be inspected carefully. Check preloading of cables used for internal bracing. Preloading of cables in a frame should be done simultaneously to prevent distortion.

FW with traffic openings should be in accordance with approved FW plans and specifications. In addition, adequacy of vertical and horizontal clearance as shown on the FW plans and as specified in the project specifications should be checked. Inspect the FW lighting system to ensure that the system is detailed according to the approved FW lighting plans.

During concrete pour, inspect FW for the following: excessive settlement, crushing of wedges, and deflection of bracing or distortion of its connection.

4.6.2 Shoring, Sloping, and Benching

Shoring, sloping, and benching are methods to prevent excavation cave-ins. Application of these methods depends on soil type, depth, and size of excavation.

Shoring systems are used to support the sides of excavations from cave-in. Steel members, timber members, or a combination of both are used to construct shoring systems. Trenching is similar to shoring, but the excavation is narrow relative to its length; the width at bottom is less than 5 m. During excavation, verify that the soil properties are the same as was anticipated in the design. Make sure that shoring members have the size and spacing as shown in the approved plans. If a tieback system is used, cables should be preloaded. Inspections should be made after a rain storm or other hazardous conditions. The inspector should ensure that ladders, ramps, or other safe means are provided in excavated areas for providing safe access.

4.6.3 Guying Systems

Guying systems are temporary structures used to stabilize column cages during construction. Guying systems usually consist of a set of cables connecting the cage or the form to heavy loads such as deadmen or K-rail. This system is designed to resist an assumed wind load applied to the cage. The inspector should check the guying system calculations submitted by the contractor for the adequacy of the system and also to ensure that the erection sequence and the removing of guys to place forms is performed in accordance with the approved plans.

4.6.4 Concrete Forms

Concrete forms should be mortar-tight and with sufficient strength to prevent excessive deflection during placement of concrete. Forms are used to hold the concrete in its plastic state until it is hardened. Forms should be cleaned of all dirt, mortar, debris, nails, and wires. Forms that will later be removed should be thoroughly coated with form oil prior to use (Figure 4.11). Forms should be wet down before placing concrete.

FIGURE 4.11 Temporary deck form for steel bridge.

For foundations, attention should be given to bracing of forms to prevent any movement during concrete placement. The bottom of forms should be checked for gaps that may cause excessive leakage of concrete.

4.7 Safety

The primary responsibility of the inspector is to ensure that a safe working environment and practices are maintained at the project site. They should set an example by following the code of safe practice and also by using personal safety equipment including hard hats, gloves, and protective clothing. In addition, they must enforce the safety issues as specified in the contract specifications. This may involve monitoring the operation of equipment and other construction equipment including barricades, warning lights, and reflectors to ensure that they are installed in accordance with the plans and specifications [22].

Prior to entering elevated or excavated areas, the inspector should ascertain that safe access is provided and proper worker protection is in place.

4.8 Record Keeping and As-Built Plans

In addition to construction inspection, the inspector is also responsible for maintaining an accurate and complete record of work that is being performed by contractors. Project records and reports are necessary to determine that contract requirements have been met so that payments can be made to the contractor. Project records should be kept current, complete, accurate, and should be submitted on time.

It is critical that the inspector keep a written diary of the activities that take place in the field. The diary should contain information concerning the work being inspected, including unusual incidents and important conversations. This information may become very critical in case of legal action for litigation involving construction claims or job failure.

As-built plans should reflect any deviation that may exist between the project plans and what was built in the field. Accurate and complete as-built plans are very important and useful for maintaining the bridge, and any future work on the bridge. The as-built plans also provide input and information for future seismic retrofit of the bridge.

4.9 Summary

This chapter discusses the construction inspection of new bridges. It provided guidelines for inspecting the main materials commonly used in building bridges and the major construction operations. Although more emphasis is placed on typical short- to medium-length bridges, the same principles are applied to other type of bridges.

References

1. Caltrans, *Standard Specifications*, California Department of Transportation, Sacramento, 1995.
2. Caltrans, *Standard Plans*, California Department of Transportation, Sacramento, 1995.
3. AASHTO, *LRFD Bridge Design Specifications*, American Association of State Highway and Transportation Officials, Washington, D.C., 1994.
4. AWS, *Bridge Welding Code*, ANSI/AASHTO/AWS D1.5-95, American Welding Society, Miami, FL, 1995.
5. Caltrans, *Falsework Manual*, Office of Structure Construction, California Department of Transportation, Sacramento, 1988.
6. Caltrans, *Trenching & Shoring Manual*, Office of Structure Construction, California Department of Transportation, Sacramento, 1990.
7. FHWA, Lateral Support Systems and Underpinning, Vol. 1: Design and Construction, Report No. FHWA-RD-75-128, U.S. Federal Highway Administration, Washington, D.C., 1976.

8. Mehtlan, J., *Outline of Field Construction Procedure*, Office of Structure Construction, California Department of Transportation, Sacramento, 1988.

9. Barsom, J. M., Properties of bridge steels, Vol. I, Chap. 3, in *Highway Structures Design Handbook*, American Institute of Steel Construction, Chicago, 1994.

10. Caltrans, *Bridge Construction Survey Guide*, Office of Structure Construction, California Department of Transportation, Sacramento, 1991.

11. AWS, *Welding Inspection*, 2nd ed., American Welding Society, Miami, FL, 1980.

12. AWS, *Guide for the Visual Inspection of Welds*, ANSI/AWS B1.11-88, American Welding Society, Miami, FL, 1997.

13. FHWA, High-Strength Bolts for Bridges, FHWA-SA-91-031, U.S. Department of Transportation, Washington, D.C., 1991.

14. AISC, Mechanical fasteners for steel bridges, Vol. I, Chap. 4A, in *Highway Structures Design Handbook*, American Institute of Steel Construction, Chicago, 1996.

15. Caltrans, *Bridge Construction Record & Procedures*, California Department of Transportation, Sacramento, 1994.

16. Caltrans, *Foundation Manual*, Office of Structure Construction, California Department of Transportation, Sacramento, 1996.

17. Caltrans, *Prestress Manual*, California Department of Transportation, Sacramento, 1992.

18. AISC, Steel erection for highway, railroad and other bridge structures, Vol. I, Chap. 14, in *Highway Structures Design Handbook*, American Institute of Steel Construction, Chicago, 1994.

19. Caltrans, *Source Inspection Manual*, Office of Materials Engineering and Testing Services, California Department of Transportation, Sacramento, 1995.

20. TxDOT, *Bridge Construction Inspection*, Texas Department of Transportation, Austin, TX, 1997.

21. Caltrans, *Bridge Deck Construction Manual*, Office of Structure Construction, California Department of Transportation, Sacramento, 1991.

22. FHWA, Bridge Inspector's Training Manual/90, FHWA-PD-91-015, U.S. Department of Transportation, Washington, D.C., 1991.

5

Maintenance Inspection and Rating

Murugesu
Vinayagamoorthy
*California Department
 of Transportation*

5.1 Introduction

Before the 1960s, little emphasis was given to inspection and maintenance of bridges in the United States. After the 1967 tragic collapse of the Silver Bridge at Point Pleasant in West Virginia, national interest in inspection and maintenance rose considerably. The U.S. Congress passed the Federal Highway Act of 1968 which resulted in the establishment of the National Bridge Inspection Standard (NBIS). The NBIS sets the national policy regarding bridge inspection procedure, inspection frequency, inspector qualifications, reporting format, and rating procedures. In addition to the establishment of NBIS, three manuals — FHWA Bridge Inspector's Training Manual 70 [1], AASHO *Manual for Maintenance Inspection of Bridges* [2], and FHWA Recording and Coding Guide for the Structure Inventory and Appraisal of the Nation's Bridges [3] — have been developed and updated [4–10] since the 1970s. These manuals along with the NBIS provide definitive guidelines for bridge inspection. Over the past three decades, the bridge inspection program evolved into one of the most sophisticated bridge management systems. This chapter will focus only on the basic, fundamental requirements for maintenance inspection and rating.

5.2   Maintenance Documentation

Each bridge document needs to have items such as structure information, structural data and history, description on and below the structure, traffic information, load rating, condition and appraisal ratings, and inspection findings. The inspection findings should have the signature of the inspection team leader.

All states in the United States are encouraged, but not mandated, to use the codes and instructions given in the Recording and Coding Guide [8,9] while documenting the bridge inventory. In order to maintain the nation's bridge inventory, FHWA requests all state agencies to submit data on the Structure Inventory and Appraisal (SI&A) Sheet. The SI&A sheet is a tabulation of pertinent information about an individual bridge. The information on SI&A sheet is a valuable aid to establish maintenance and replacement priorities and to determine the maintenance cost of the nation's bridges.

5.3   Fundamentals of Bridge Inspection

5.3.1   Qualifications and Responsibilities of Bridge Inspectors

The primary purpose of bridge inspection is to maintain the public safety, confidence, and investment in bridges. Ensuring public safety and investment decisions require a comprehensive bridge inspection. To this end, a bridge inspector should be knowledgeable in material and structural behavior, bridge design, and typical construction practices. In addition, inspectors should be physically strong because the inspection sometimes requires climbing on rough, steep, and slippery terrain, working at heights, or working for days.

Some of the major responsibilities of a bridge inspector are as follows:

- Identifying minor problems that can be corrected before they develop into major repairs;
- Identifying bridge components that require repairs in order to avoid total replacement;
- Identifying unsafe conditions;
- Preparing accurate inspection records, documents, and recommendation of corrective actions; and
- Providing bridge inspection program support.

In the United States, NBIS requires a field leader for highway bridge inspection teams. The field team leader should be either a professional engineer or a state certified bridge inspector, or a Level III bridge inspector certified through the National Institute for Certification of Engineering Technologies. It is the responsibility of the inspection team leader to decide the capability of individual team members and delegate their responsibilities accordingly. In addition, the team leader is responsible for the safety of the inspection team and establishing the frequency of bridge inspections.

5.3.2   Frequency of Inspection

NBIS requires that each bridge that is opened to public be inspected at regular intervals not exceeding 2 years. The underwater components that cannot be visually evaluated during periods of low flow or examined by feel for their physical conditions should be inspected at an interval not exceeding 5 years.

The frequency, scope, and depth of the inspection of bridges generally depend on several parameters such as age, traffic characteristics, state of maintenance, fatigue-prone details, weight limit posting level, and known deficiencies. Bridge owners may establish the specific frequency of inspection based on the above factors.

5.3.3   Tools for Inspection

In order to perform an accurate and comprehensive inspection, proper tools must be available. As a minimum, an inspector needs to have a 2-m (6-ft) pocket tape, a 30-m (100-ft) tape, a chipping hammer, scrapers, flat-bladed screwdriver, pocketknife, wire brush, field marking crayon, flashlight, plumb bob, binoculars, thermometer, tool belt with tool pouch, and a carrying bag. Other useful tools are a shovel, vernier or jaw-type calipers, lighted magnifying glass, inspection mirrors, dye penetrant, 1-m (4-ft) carpenter's level, optical crack gauge, paint film gauge, and first-aid kits. Additional special inspection tools are survey, nondestructive testing, and underwater inspection equipment.

Inspection of a bridge prompts several unique challenges to bridge inspectors. One of the challenges to inspectors is the accessibility of bridge components. Most smaller bridges can be accessed from below without great effort, but larger bridges need the assistance of accessing equipment and vehicles. Common access equipment are ladders, rigging, boats or barges, floats, and scaffolds. Common access vehicles are manlifts, snoopers, aerial buckets, and traffic protection devices. Whenever possible, it is recommended to access the bridge from below since this eliminates the need for traffic control on the bridge. Setting up traffic control may create several problems, such as inconvenience to the public, inspection cost, and safety of the public and inspectors.

5.3.4   Safety during Inspection

During the bridge inspection, the safety of inspectors and of the public using the bridge or passing beneath the bridge should be given utmost importance. Any accident can cause pain, suffering, permanent disability, family hardship, and even death. Thus, during the inspection, inspectors are encouraged to follow the standard safety guidelines strictly.

The inspection team leader is responsible for creating a safe environment for inspectors and the public. Inspectors are always encouraged to work in pairs. As a minimum, inspectors must wear safety vests, hard hats, work gloves, steel-toed boots, long-sleeved shirts, and long pants to ensure their personal safety. Other safety equipment are safety goggles, life jackets, respirator, gloves, and safety belt. A few other miscellaneous safety items include walkie-talkies, carbon monoxide detectors, and handheld radios.

Field clothes should be appropriate for the climate and the surroundings of the inspection location. When working in a wooded area, appropriate clothing should be worn to protect against poisonous plants, snakes, and disease-carrying ticks. Inspectors should also keep a watchful eye for potential hazardous environments around the inspection location. When entering a closed bridge box cell, air needs to be checked for the presence of oxygen and toxic or explosive gases. In addition, care should be taken when using existing access ladders and walkways since the ladder rungs may be rusted or broken. When access vehicles such as snoopers, booms, or rigging are used, the safe use of this equipment should be reviewed before the start of work.

5.3.5   Reports of Inspection

Inspection reports are required to establish and maintain a bridge history file. These reports are useful in identifying and assessing the repair requirements and maintenance needs of bridges. NBIS requires that the findings and results of a bridge inspection be recorded on standard inspection forms. Actual field notes and numerical conditions and appraisal ratings should be included in the standard inspection form. It is also important to recognize that these inspection reports are legal documents and could be used in future litigation.

Descriptions in the inspection reports should be specific, detailed, quantitative, and complete. Narrative descriptions of all signs of distress, failure, or defects with sufficient accuracy should be noted so that another inspector can make a comparison of condition or rate of disintegration in the future. One example of a poor description is, "Deck is in poor condition." A better description would be, "Deck is in poor condition with several medium to large cracks and numerous spalls." The seriousness and the amount of all deficiencies must be clearly stated in an inspection report.

In addition to inspection findings about the various bridge components, other important items to be included in the report are any load, speed, or traffic restrictions on the bridge; unusual loadings; high water marks; clearance diagram; channel profile; and work or repairs done to the bridge since the last inspection.

When some improvement or maintenance work alters the dimensions of the structure, new dimensions should be obtained and reported. When the structure plans are not in the history file, it may be necessary to prepare plans using field measurements. These measurements will later be used to perform the rating analysis of the structure.

Photographs and sketches are the most effective ways of describing a defect or the condition of structural elements. It is therefore recommended to include sketches and/or photographs to describe or illustrate a defect in a structural element. At least two photographs for each bridge for the record are recommended.

Other tips on photographs are

- Place some recognizable items that will allow the reviewer to visualize the scale of the detail;
- Include plumb bob to show the vertical line; and
- Include surrounding details so one could relate other details with the specific detail.

After inspecting a bridge, the inspector should come to a reasonable conclusion. When the inspector cannot interpret the inspection findings and determine the cause of a specific finding (defect), the advice of more-experienced personnel should be sought. Based on the conclusion, the inspector may need to make a practical recommendation to correct or preclude bridge defects or deficiencies. All instructions for maintenance work, stress analysis, posting, further inspection, and repairs should be included in the recommendation. Whenever recommendations call for bridge repairs, the inspector must carefully describe the type of repairs, the scope of the work, and an estimate of the quantity of materials.

5.4   Inspection Guidelines

5.4.1   Timber Members

Common damage in timber members is caused by fungi, parasites, and chemical attack. Deterioration of timber can also be caused by fire, impact or collisions, abrasion or mechanical wear, overstress, and weathering or warping.

Timber members can be inspected by both visual and physical examination. Visual examination can detect the following: fungus decay, damage by parasites, excessive deflection, checks, splits, shakes, and loose connections. Once the damages are detected visually, the inspector should investigate the extent of each damage and properly document them in the inspection report. Deterioration of timber can also be detected using sounding methods — a nondestructive testing method. Tapping on the outside surface of the member with a hammer detects hollow areas, indicating internal decay. There are a few advanced nondestructive and destructive techniques available. Two of the commonly used destructive tests are boring or drilling and probing. And, two of the nondestructive tests are Pol-Tek and ultrasonic testing. The Pol-Tek method is used to detect low-density regions and ultrasonic testing is used to measure crack and flaw size.

5.4.2   Concrete Members

Common concrete member defects include cracking, scaling, delamination, spalling, efflorescence, pop-outs, wear or abrasion, collision damage, scour, and overload. Brief descriptions of common damages are given in this section.

Cracking in concrete is usually large enough to be seen with the naked eye, but it is recommended to use a crack gauge to measure and classify the cracks. Cracks are classified as hairline, medium, or wide cracks. Hairline cracks cannot be measured by simple means such as pocket ruler, but simple means can be used for the medium and wide cracks. Hairline cracks are usually insignificant to the capacity of the structure, but it is advisable to document them. Medium and wide cracks are significant to the structural capacity and should be recorded and monitored in the inspection reports. Cracks can also be grouped into two types: structural cracks and nonstructural cracks. Structural cracks are caused by the dead- and live-load stresses.

Structural cracks need immediate attention, since they affect the safety of the bridge. Nonstructural cracks are usually caused by thermal expansion and shrinkage of the concrete. These cracks are insignificant to the capacity, but these cracks may lead to serious maintenance problems. For example, thermal cracks in a deck surface may allow water to enter the deck concrete and corrode the reinforcing steel.

Scaling is the gradual and continuing loss of surface mortar and aggregate over an area. Scaling is classified into four categories: light, medium, heavy, and severe.

Delamination occurs when layers of concrete separate at or near the level of the top or outermost layer of reinforcing steel. The major cause of delamination is the expansion or the corrosion of reinforcing steel due to the intrusion of chlorides or salts. Delaminated areas give off a hollow sound when tapped with a hammer. When a delaminated area completely separates from the member, a roughly circular or oval depression, which is termed a spall, will be formed in the concrete.

The inspection of concrete should include both visual and physical examination. Two of the primary deteriorations noted by visual inspections are cracks and rust stains. An inspector should recognize the fact that not all cracks are of equal importance. For example, a crack in a prestressed concrete girder beam, which allows water to enter the beam, is much more serious than a vertical crack in the backwall. A rust stain on the concrete members is one of the signs of corroding reinforcing steel in the concrete member. Corroded reinforcing steel produces loss of strength within concrete due to reduced reinforced steel section, and loss of bond between concrete and reinforcing steel. The length, direction, location, and extent of the cracks and rust stains should be measured and reported in the inspection notes.

Some common types of physical examination are hammer sounding and chain drag. Hammer sounding is used to detect areas of unsound concrete and usually used to detect delaminations. Tapping the surfaces of a concrete member with a hammer produces a resonant sound that can be used to indicate concrete integrity. Areas of delamination can be determined by listening for hollow sounds. The hammer sounding method is impractical for the evaluation of larger surface areas. For larger surface areas, chain drag can be used to evaluate the integrity of the concrete with reasonable accuracy. Chain drag surveys of decks are not totally accurate, but they are quick and inexpensive.

There are other advanced techniques — destructive and nondestructive — available for concrete inspection. Core sampling is one of the destructive techniques of concrete inspection. Some of the nondestructive inspection techniques are

- Delamination detection machinery to identify the delaminated deck surface;
- Copper sulfate electrode, nuclear methods to determine corrosion activity;
- Ground-penetrating radar, infrared thermography to detect deck deterioration;
- Pachometer to determine the position of reinforcement; and
- Rebound and penetration method to predict concrete strength.

5.4.3   Steel and Iron Members

Common steel and iron member defects include corrosion, cracks, collision damage, and overstress. Cracks usually initiate at the connection detail, at the termination end of a weld, or at a corroded location of a member and then propagate across the section until the member fractures. Since all of the cracks may lead to failure, bridge inspectors need to look at each and every one of these potential crack locations carefully. Dirt and debris usually form on the steel surface and shield the defects on the steel surface from the naked eye. Thus, the inspector should remove all dirt and debris from the metal surface, especially from the surface of fracture-critical details, during the inspection of defects.

The most recognizable type of steel deterioration is corrosion. The cause, location, and extent of the corrosion need to be recorded. This information can be used for rating analysis of the member and for taking preventive measures to minimize further deterioration. Section loss due to corrosion can be reported as a percentage of the original cross section of a component. The corrosion section loss is calculated by multiplying the width of the member and the depth of the defect. The depth of the defect can be measured using a straightedge ruler or caliper.

One of the important types of damage in steel members is fatigue cracking. Fatigue cracks develop in bridge structures due to repeated loadings. Since this type of cracking can lead to sudden and catastrophic failure, the bridge inspector should identify fatigue-prone details and should perform a thorough inspection of these details. For painted structures, breaks in the paint accompanied by rust staining indicate the possible existence of a fatigue crack. If a crack is suspected, the area should be cleaned and given a close-up visual inspection. Additionally, further testing such as dye penetrant can be done to identify the crack and to determine its extent. If fatigue cracks are discovered, inspection of all similar fatigue details is recommended.

Other types of damage may occur due to overstress, vehicular collision, and fire. Symptoms of damage due to overstress are inelastic elongation (yielding) or decrease in cross section (necking) in tension members, and buckling in compression members. The causes of the overstress should be investigated. The overstress of a member could be the result of several factors such as loss of composite action, loss of bracing, loss of proper load-carrying path, and failure or settlement of bearing details.

Damage due to vehicular collision includes section loss, cracking, and shape distortion. These types of damage should be carefully documented and repair work process should be initiated. Until the repair work is completed, restriction of vehicular traffic based on the rating analysis results is recommended.

Similar to timber and concrete members, there are advanced destructive and nondestructive techniques available for steel inspection. Some of the nondestructive techniques used in steel bridges are

- Acoustic emissions testing to identify growing cracks;
- Computer tomography to render the interior defects;
- Dye penetrant to define the size of the surface flaws; and
- Ultrasonic testing to detect cracks in flat and smooth members.

5.4.4   Fracture-Critical Members

Fracture-critical members (FCM) or member components are defined as tension components of members whose failure would be expected to result in collapse of a portion of a bridge or an entire bridge [7,8]. A redundant steel bridge that has multiple load-carrying mechanisms is seldom categorized as a fracture-critical bridge.

Since the failure to locate defects on FCMs in a timely manner may lead to catastrophic failure of a bridge, it is important to ensure that FCMs are inspected thoroughly. Hands-on involvement of the team leader is necessary to maintain the proper level of inspection and to make independent checks of condition appraisals. In addition, adequate time to conduct a thorough inspection should be allocated by the team leader. Serious problems in FCMs must be addressed immediately by restricting traffic on the bridge and repairing the defects under an emergency contract. Less serious problems requiring repairs or retrofit should be placed on the programmed repair work so that they will be incorporated into the maintenance schedule.

Bridge inspectors need to identify the FCMs using the guidelines provided in the Inspection of Fracture Critical Bridge Members [7,8]. There are several vulnerable fracture-critical locations in a bridge. Some of the obvious locations are field welds, nonuniform welds, welds with unusual profile, and intermittent welds along the girder. Other possible locations are insert plate termination points, floor beam to girder connections, diaphragm connection plates, web stiffeners, areas that are vulnerable to corrosion, intersecting weld location, sudden change in cross section, and coped sections. Detailed descriptions of each of these fracture-critical details are listed in the Inspection of Fracture Critical Bridge Members [7,8]. Once the FCM is identified in a bridge structure, information such as location, member components, and likelihood to have fatigue- or corrosion-related damage needs to be gathered. The information gathered on the member should become a permanent record and the condition of the member should be updated on every subsequent inspection.

FCMs can be inspected by both visual and physical examination. During the visual inspection, the inspector performs a close-up, hands-on inspection using standard, readily available tools. During the physical examination, the inspector uses the most-sophisticated nondestructive testing methods. Some of the FCMs may have details that are susceptible to fatigue cracking and others may be in poor condition due to corrosion.

5.4.5 Scour-Critical Bridges

Bridges spanning over waterways, especially rivers and streams, sometimes provide major maintenance challenges. These bridges are susceptible to scour of the riverbed. When the scoured riverbed elevation falls below the top of the footing, the bridge is referred to as scour critical.

The rivers, whether small or large, could significantly change their size over the period of the lifetime of a bridge. A riverbed could be altered in several ways and thereby jeopardize the stability of the bridges. A few of the possible types of riverbed alterations are scour, hydraulic opening, channel misalignment, and bank erosion. Scour around the bridge substructures poses potential structural stability concerns. Scour at bridges depends on the hydraulic features upstream and downstream, riverbed sediments, substructure section profile, shoreline vegetation, flow velocities, and potential debris. The estimation of the overall scour depth will be used to identify scour-prone and scour-critical bridges. Guidance for the scour evaluation process is provided in Evaluating Scour at Bridges [11].

A typical scour evaluation process falls into two phases: inventory phase and evaluation phase. The main goal of the inventory phase is to identify those bridges that are vulnerable to scour (scour-prone bridges). Evaluation during this phase is made using the available bridge records, inspection records, history of the bridge, original stream location, evidence of scour, deposition of debris, geology, and general stability of the streambed. Once the scour-prone bridges are identified, the evaluation phase needs to be performed. The scour evaluation phase requires in-depth field review to generate data for estimation of the hydraulics and scour depth. The procedure of scour estimation is outlined in Evaluating Scour at Bridges [11]. The scour depths are then compared with the existing foundation condition. When the scour depth is above the top of the footing, the bridge would require no action. However, when the scour depth is within the limits of the footing or piles, a structural stability analysis is needed. If the scour depth is below the pile tips or spread footing base, monitoring of the bridge is required. The results obtained from the scour evaluation process are entered into the bridge inventory.

5.4.6 Underwater Components

Underwater components are mostly substructure members. Since the accessibility of these members is difficult, special equipment is necessary to inspect these underwater components. Also, visibility during the underwater inspection is generally poor, and therefore a thorough inspection of the members will not be possible. Underwater inspection is classified as visual (Level 1), detailed (Level 2), and comprehensive (Level 3) to specify the level of effort of inspection. Details of these various levels of inspection are discussed in the *AASHO Manual for Maintenance Inspection of Bridges* [2] and Evaluating Scour at Bridges [11].

Underwater steel structure components are susceptible to corrosion, especially in the low to high water zone. Some of the defects observed in underwater timber piles are splitting, decay or rot, marine borers, decay of timber at connections, and corrosion of connectors. It is important to recognize that the timber piles may appear sound on the outside shell but be severely damaged inside. Some of the most common defects in underwater concrete piles are cracking, spalls, exposed reinforcing, sulfate attack, honeycombing, and scaling. When cracking, spalls, and exposed reinforcing are detected, structural analysis may be required to ensure the safety of the bridge.

5.4.7 Decks

The materials typically used in the bridge structures are concrete, timber, and steel. Previous sections discuss some of the defects associated with each of these materials. In this section, the damage most likely to occur in bridge decks is discussed.

Common defects in steel decks are cracked welds, broken fasteners, corrosion, and broken connections. In a corrugated steel flooring system, section loss due to corrosion may affect the load-carrying capacity of the deck and thus the actual amount of remaining materials needs to be evaluated and documented.

Common defects in timber decks are crushing of the timber deck at the supporting floor system, flexure damages such as splitting, sagging, and cracks in tension areas, and decay of the deck due to biological organisms, especially in the areas exposed to drainage.

Common defects in concrete decks are wear, scaling, delamination, spalls, longitudinal flexure cracks, transverse flexure cracks in the negative moment regions, corrosion of the deck rebars, cracks due to reactive aggregates, and damage due to chemical contamination. The importance of a crack varies with the type of concrete deck. A large to medium crack in a noncomposite deck may not affect the load-carrying capacity of the main load-carrying member. On the other hand, several cracks in a composite deck will affect the structural capacity. Thus, an inspector must be able to identify the functions of the deck while inspecting it.

Sometimes a layer of asphalt concrete (AC) overlay will be placed to provide a smooth driving and wearing surface. Extra care is needed during the inspection, because AC overlay prevents the inspector's ability to inspect the top surface of the deck visually for damage.

5.4.8 Joint Seals

Damage to the joint seals is caused by vehicle impact, extreme temperature, and accumulation of dirt and debris. Damage from debris and vehicles such as snowplows could cause the joint seals to be torn, pulled out of anchorage, or removed altogether. Damage from extreme temperature could break the bond between the joint seal and deck and consequently result in pulling out the joint seal altogether.

The primary function of deck joints is to accommodate the expansion and contraction of the bridge superstructure. These deck joints also provide a smooth transition from the approach roadway to the bridge deck. Deck joints are placed at hinges between two decks of adjacent structures, and between the deck sections and abutment backwall. The joint seals used in the bridge industry can be divided into two groups: open joints and closed joints. Open joints allow water and debris to pass through the joints. Dripping water through open joints usually damages the bearing details. Closed joints do not allow water and debris to pass through them. A few of the closed joints are compression seal, poured joint seal, sliding plate joint, plank seal, sheet seal, and strip seal.

In the case of closed joints, damage to the joint seal material will cause the water to drip on the bearing seats and consequently damage the bearing. Accumulation of dirt and debris may prevent normal thermal expansion and contraction, which may in turn cause cracking in the deck, backwall, or both. Cracking in the deck may affect the ride quality of the bridge, may produce larger impact load from vehicles, and may reduce the live-load-carrying capacity of the bridge.

5.4.9 Bearings

Bearings used in bridge structures could be categorized into two groups: metal and elastomeric. Metal bearings sometimes become inoperable (sometimes referred as "frozen") due to corrosion, mechanical bindings, buildup of debris, or other interference. Frozen bearings may result in bending, buckling, and improper alignment of members. Other types of damage are missing fasteners, cracked welds, corrosion on the sliding surface, sole plate rests only on a portion of the masonry plate, and binding of lateral shear keys.

Damage in elastomeric bearing pads is excessive bulging, splitting or tearing, shearing, and failure of bond between sole and masonry plate. Excessive bulging indicates that the bearing might be too tall. When the pad is under excessive strain for a long period, the pad will experience shearing failure.

Inspectors need to assess the exact condition of the bearing details and to recommend corrective measures that allow the bearing details to function properly. Since the damage to the bearings will affect the other structural members as time passes, repair of bearing damage needs to be considered as a preventive measure.

5.5 Fundamentals of Bridge Rating

5.5.1 Introduction

Once a bridge is constructed, it becomes the property of the owner or agency. The evaluation or rating of existing bridges is a continuous activity of the agency to ensure the safety of the public. The evaluation provides necessary information to repair, rehabilitate, post, close, or replace the existing bridge.

In the United States, since highway bridges are designed for the AASHTO design vehicles, most U.S. engineers tend to believe that the bridge will have adequate capacity to handle the actual present traffic. This belief is generally true if the bridge was constructed and maintained as shown in the design plan. However, changes in a few details during the construction phase, failure to attain the recommended concrete strength, unexpected settlements of the foundation after construction, and unforeseen damage to a member could influence the capacity of the bridge. In addition, old bridges might have been designed for a lighter vehicle than is used at present, or a different design code. Also, the live-load-carrying capacity of the bridge structure may have altered as a result of deterioration, damage to its members, aging, added dead loads, settlement of bents, or modification to the structural member.

Sometimes, an industry would like to transport their heavy machinery from one location to another location. These vehicles would weigh much more than the design vehicles and thus the bridge owner may need to determine the current live-load-carrying capacity of the bridge. In the following sections, establishing the live load-carrying capacity and the bridge rating will be discussed.

5.5.2 Rating Principles

In general, the resistance of a structural member (R) should be greater than the demand (Q) as follows:

$$R \geq Q_d + Q_l + \sum_i Q_i \tag{5.1}$$

where Q_d is the effect of dead load, Q_l is the effect of live load, and Q_i is the effect of load i.

Eq. (5.1) applies to design as well as evaluation. In the bridge evaluation process, maximum allowable live load needs to be determined. After rearranging the above equation, the maximum allowable live load will become

$$Q_l \leq R - \left(Q_d + \sum_i Q_i \right) \tag{5.2}$$

Maintenance engineers always question whether a fully loaded vehicle (rating vehicle) can be allowed on the bridge and, if not, what portion of the rating vehicle could be allowed on a bridge. The portion of the rating vehicle will be given by the ratio between the available capacity for live-load effect and the effect of the rating vehicle. This ratio is called the rating factor (RF).

$$\text{RF} = \frac{\text{Available capacity for the live-load effect}}{\text{Rating vehicle load demand}} = \frac{R - \left(Q_d + \sum_i Q_i \right)}{Q_l} \tag{5.3}$$

When the rating factor equals or exceeds unity, the bridge is capable of carrying the rating vehicle. On the other hand, when the rating factor is less than unity the bridge may be overstressed while carrying the rating vehicle.

The capacity of a member is usually independent of the live-load demand. Thus, Eq. (5.3) is generally a linear expression. However, there are cases where the capacity of a member is dependent on the live-load forces. For example, available moment capacity depends on the total axial load in biaxial bending members. In a biaxially loaded member, Eq. (5.3) will be a second-order expression.

Thermal, wind, and hydraulic loads may be neglected in the evaluation process because the likelihood of occurrence of extreme values during the relatively short live-load loading is small. Thus, the effects of the dead and live loads are the only two loads considered in the evaluation process.

5.5.3   Rating Philosophies

During the structural evaluation process, the location and type of critical failure modes are first identified; Eq. (5.3) is then solved for each of these potential failures. Although the concept of evaluation is the same, the mathematical relationship of this basic equation for allowable stress design (ASD), load factor design (LFD), and load and resistance factor design (LRFD) differs. Since the resistance and load effect can never be established with certainty, engineers use safety factors to give adequate assurance against failure. ASD includes safety factors in the form of allowable stresses of the material. LFD considers the safety factors in the form of load factors to account for the uncertainty of the loadings and resistance factors to account for the uncertainty of structural response. LRFD treats safety factors in the form of load and resistance factors that are based on the probability of the loadings and resistances.

For ASD, the rating factor expression Eq. (5.3) can be written as

$$RF = \frac{R - \left(\sum D + \sum_i L_i(1+I) \right)}{L(1+I)} \tag{5.4}$$

For LFD, the rating factor expression Eq. (5.3) can be written as

$$RF = \frac{\phi R_n - \sum \gamma_D D - \sum_{i=1}^{n} \gamma_{Li} L_i(1+I)}{\gamma_L L(1+I)} \tag{5.5}$$

For LFRD, the rating factor expression Eq. (5.3) can be written as

$$RF = \frac{\phi R_n - \sum \gamma_D D - \sum_{i=1}^{n} \gamma_{Li} L_i(1+I)}{\gamma_L L(1+I)} \tag{5.6}$$

where R is the allowable stress of the member; ϕR_n is nominal resistance; D is the effect of dead loads; L_i is the live-load effect for load i other than the rating vehicle; L the nominal live-load effect of the rating vehicle; I is the impact factor for the live-load effect; γ_D, γ_{Li}, and γ_L are dead- and live-load factors, respectively.

Researchers are now addressing the LRFD method, and thus the LRFD approach may be revised in the near future. Since the LRFD method is being developed at this time, the LRFD method is not discussed further in this chapter.

In order to use the above equations (Eqs. 5.4 to 5.6) in determining the rating factors, one needs to estimate the effects of individual live-load vehicles. The effect of individual live-load vehicles on structural member could only be obtained by analyzing the bridge using a three-dimensional analysis. Thus, obtaining the rating factor using the above expressions is very difficult and time-consuming.

To simplify the above equations, it is assumed that similar rating vehicles will occupy all the possible lanes to produce the maximum effect on the structure. This assumption allows us to use the AASHTO live-load distribution factor approach to estimate the live-load demand and eliminate the need for the three-dimensional analysis.

And the simplified rating factor equations become as follows:

$$\text{For ASD: RF} = \frac{R - D}{L(1 + I)} \tag{5.7}$$

$$\text{For LFD: RF} = \frac{\phi R_n - \gamma_D D}{\gamma_L L(1 + I)} \tag{5.8}$$

$$\text{For LRFD: RF} = \frac{\phi R_n - \gamma_D D}{\gamma_L L(1 + I)} \tag{5.9}$$

In the derivation of the above equations (Eqs. 5.7 to 5.9), it is assumed that the resistance of the member is independent of the loads. A few exceptions to this assumption are beam–column members and beams with high moment and shear. In a beam–column member, axial capacity or moment capacity depends on the applied moment or applied axial load on the member. Thus, as the live-load forces in the member increase, the capacity of the member would decrease. In other words, the numerator of the above equations (available live-load capacity) will drop as the live load increases. Thus, the rating factor will no longer be a constant value, and will be a function of live load.

5.5.4   Level of Ratings

There are two levels of rating for bridges: inventory and operating. The rating that reflects the absolute maximum permissible load that can be safely carried by the bridge is called an operating rating. The load that can be safely carried by a bridge for an indefinite period is called an inventory rating.

The life of a bridge depends on the fatigue life or serviceability limits of bridge materials. Higher frequent loading and unloading may affect the fatigue life or serviceability of a bridge component and thereby the life of the bridge. Thus, in order to maintain a bridge for an indefinite period, live-load-carrying capacity available for frequently passing vehicles needs to be estimated at service. This process is referred to as inventory rating.

Less frequent vehicles may not affect the fatigue life or serviceability of a bridge, and thus live-load-carrying capacity available for less frequent vehicles need not be estimated using serviceability criteria. In addition, since less frequent vehicles do not damage the bridge structure, bridge structures could be allowed to carry higher loads. This process is referred to as operating rating.

5.5.5   Structural Failure Modes

In the ASD approach, when a portion of a structural member is stressed beyond the allowable stress, the structure is considered failed. In addition, since any portion of the structural member material never reaches its yield, the deflections or vibrations will always be satisfied. Thus, the serviceability of a bridge is ensured when the allowable stress method is used to check a bridge member. In other words, in the ASD approach, serviceability and strength criteria are satisfied automatically. The inventory and operating allowable stresses for various types of failure modes are given in the AASHTO *Manual for Condition Evaluation of Bridges 1994* [12].

In the LFD approach, failure could occur at two different limit states: serviceability and strength. When the load on a member reaches the ultimate capacity of the member, the structure is considered failed at its ultimate strength limit state. When the structure reaches its maximum allowable serviceability limits, the structure is considered failed at its serviceability limit state. In the LFD approach, satisfying one of the limit states will not automatically guarantee the satisfaction of the other limit state. Thus, both serviceability and strength criteria need to be checked in the LFD method. However, when the operating rating is estimated, the serviceability limits need not be checked.

5.6 Superstructure Rating Examples

In this section, several problems are illustrated to show the bridge rating procedures. In the following examples, AASHTO *Standard Specification for Highway Bridges*, 16th ed. [13] is referred to as Design Specifications and AASHTO *Manual for Condition Evaluation of Bridges* 1994 [12] is referred to as Rating Manual. All the notations used in these examples are defined in either the Design Specifications or the Rating Manual.

5.6.1 Simply Supported Timber Bridge

Given

Typical cross section of a 16-ft (4.88-m) long simple-span timber bridge is shown in Figure 5.1. 13.4 × 16 in. (101.6 × 406.4 mm) timber stringers are placed at 18 in. (457 mm) spacing. 4 × 12 in. (101.6 × 305 mm) timber planks are used as decking. 8 × 8 in. (203 × 203 mm) timber is used as wheel guard. Barrier rails (10 lb/ft or 0.1 N/mm) are placed at either side of the bridge. The traffic lane width of the bridge is 16 ft (4.88 m). Assume that the allowable stresses at operating level are as follows: F_b for stringer as 1600 psi (11 MPa) and F_v of stringer level as 115 psi (0.79 MPa).

Requirement

Determine the critical rating factors for interior stringer for HS20 vehicle using the ASD approach.

Solution

For this simply supported bridge, the critical locations for ratings will be the locations where shear and moments are higher.

Shear needs to be checked at a distance (s) $3d$ or $0.25L$ from the bearing location for vehicle live loads; thus,

$$s = 3d = 3 \times 16 \text{ in.}/12 = 4.0 \text{ ft or}$$

$$= 0.25L = 0.25 \times 16 \text{ ft} = 4.0 \text{ ft.}$$

Thus, s is taken as 4.0 ft (1.22 m).

Maximum dead- and live-load shear will occur at this point and thus in the following calculations, shear is estimated at this critical location.

1. **Dead-Load Calculations**

Self-weight of the stringer	$= 0.05 \times 4 \times 16 \times \dfrac{1}{144}$	$= 0.22$ kips/ft
Weight of deck (using tributary area)	$= 1.5 \times 4 \times 12 \times \dfrac{1}{144} \times 0.05$	$= 0.025$ kips/ft
Weight of 1.5 in. AC on the deck	$= 1.5 \times \dfrac{1.5}{12} \times 0.144$	$= 0.027$ kips/ft

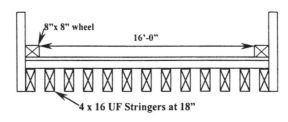

FIGURE 5.1 Typical cross section detail of simply supported timber bridge example.

Barrier rail and curb $= \left(10 + 50 \times \dfrac{8 \times 8}{144}\right) \times \dfrac{2}{13 \times 1000} = 0.004$ kips/ft

Total uniform dead load on the stringer $= 0.022 + 0.025 + 0.027 + 0.004 = 0.078$ kips/ft

Maximum dead-load moment at midspan $= \dfrac{wl^2}{8} = \dfrac{0.078 \times 16^2}{8}$ kip-ft (3390 N-m)

Maximum dead-load shear at this critical point $= w \times (0.5L - s)$

$$= 0.078 \times (0.5 \times 16 - 4)$$

$$= 0.31 \text{ kips (1.38 kN)}$$

2. Live-Load Calculations

The travel width is less than 18 ft. Thus, according to Section 6.7.2.2 of the Rating Manual, this bridge needs to be rated for one traffic lane. From Designs Specifications Table 3.23.1,

$$\text{Number of wheels on the stringer} = \frac{S}{4} = \frac{1.5}{4} = 0.38$$

Maximum moment due to HS20 loading (Appendix A3, Rating Manual)
$= (64)(0.38)$
$= 24.32$ kip-ft (33,000 N-m)
In order to estimate the live-load shear, we need to estimate the shear due to undistributed and distributed HS20 loadings. (See Design Specifications 13.6.5.2 for definition of V_{LU} and V_{LD}.)

Shear due to undistributed HS20 loadings $= V_{LU}$ $= 16 \times 12/16 = 12$ kips (53.4 kN)

Shear due to distributed HS20 loading $= V_{LD}$ $= 16 \times 12/16 \times (0.38)$

$$= 4.56 \text{ kips (20.3 kN)}$$

Thus, shear due to HS20 live load $= 0.5(0.6V_{LU} + V_{LD}) = 5.88$ kips (26.1 kN)

3. Capacity Calculations

a. *Moment capacity at midspan*:
   Moment capacity of the timber stringer at *Operating level*

$$= F_b S_x = 1600 \times \frac{1}{6} \times 4 \times 16^2 \times \frac{1}{12,000} = 22.8 \text{ kip-ft (30,900 N-m)}$$

According to Section 6.6.2.7 of Rating Manual, the operating level stress of a timber stringer can be taken as 1.33 times the inventory level stress.
Thus, moment capacity of the timber stringer at *Inventory level*

$$= 22.8/1.33$$

$$= 17.1 \text{ kip-ft (23,200 N-m)}$$

b. *Shear capacity at support*:
   Shear capacity of the timber section (controlled by horizontal shear) $= (\tfrac{2}{3})bdf_v$:

V_c at *operating level* = (2/3) × 4 × 16 × 115 psi × 1/1000 = 4.91 kips (21.8 kN)

V_c at *inventory level* = 14.91/1.33 = 3.69 kips (16.4 kN)

4. Rating Calculations

$$\text{Rating factor based on ASD method} = RF = \frac{R-D}{L(1+I)}$$

By substituting appropriate values, the rating factor can be determined.

a. *Based on moment at midspan*:

$$\text{Inventory rating factor } RF_{INV\text{-}MOM} = \frac{17.1-2.5}{24.32} = 0.600$$

$$\text{Operating rating factor } RF_{OPR\text{-}MOM} = \frac{22.8-2.5}{24.32} = 0.835$$

b. *Based on shear at the support*:

$$\text{Inventory rating factor } RF_{INV\text{-}SHE} = \frac{3.69-0.31}{5.88} = 0.575$$

$$\text{Operating rating factor } RF_{OPR\text{-}SHE} = \frac{4.91-0.31}{5.88} = 0.782$$

5. Summary

It is found that the critical rating factor is controlled by shear in the stringers. The critical inventory and operating rating of the bridge will be 0.575 and 0.782, respectively.

5.6.2   Simply Supported T-Beam Concrete Bridge

Given

A bridge, which was built in 1929, consists of three simple-span reinforced concrete T-beams on concrete bents and abutments. The span lengths are 16 ft (4.88 m), 50 ft (15.24 m), and 10 ft (3.05 m). Typical cross section and girder details are shown in Figure 5.2. General notes given in the plan indicate that f_c = 1000 psi (6.9 MPa) and f_s = 18,000 psi (124.1 MPa). Assume the weight of each barrier rail as 250 lb/ft (3.6 N/mm).

Requirement

Determine the critical rating factor of the interior girder of the second span (50 ft. or 15.24 m) for HS20 vehicles assuming no deterioration of materials occurred.

Solution

1. Dead-Load Calculations

Self-weight of the girder = (3.5) (1.333) (0.15)	= 0.700 kips/ft
(4 × 4 in.) Fillets between girder and slab = 2(1/2) (4/12) (4/12) (0.15)	= 0.017 kips/ft
Slab weight (based on tributary area) = (6.667)(8/12) (0.15)	= 0.667 kips/ft
Contribution from barrier rail (equally distributed among girders) = 2 (0.25/3)	= 0.167 kips/ft
Thus, total uniform load on the interior girder	= 1.551 kips/ft (22.6 N/mm)

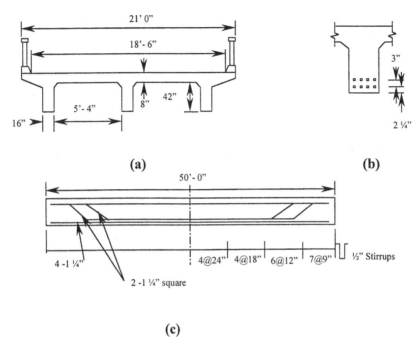

FIGURE 5.2 Details of simply supported T-beam concrete bridge example. (a) Typical cross section; (b) reinforcement locations; (c) T-beam girder details.

Dead-load moment at midspan	= 484.6 kips/ft (0.657 MN/m)
Dead-load shear at a distance d from support	= 32.31 kips (143.7 kN)

2. Live-Load Calculations

The traffic lane width of this bridge is 18.5 ft. According to Design Specifications, any bridge with a minimum traffic lane width of 18 ft needs to carry two lanes. Hence, the number of live-load wheels will be based on two traffic lanes. From Table 3.23.1A of Design Specifications for two traffic lanes for T-beams is given by $S/6.0$

Number of live-load wheel line	= 6.667/6.0	= 1.111
AASHTO standard impact factor for moment	= 50/(125 + 50)	= 0.286
AASHTO standard impact factor for shear at support	= 50/(125 + 50)	= 0.286

The live-load moments and shear tables listed in the Rating Manual are used to determine the live-load demand.

Maximum HS20 moment for 50 ft span without impact/wheel line	= 298.0 kips-ft
Thus, HS20 moment with impact at midspan = (1.286) (1.111) (298.0)	= 425.7 kips-ft (0.58 MN-m)
Maximum HS20 shear at a distance d from the support/wheel line	= 28.32 kips
Thus, maximum HS20 shear = (1.286) (1.111) (28.32)	= 40.46 kips (180.0 kN)

3. **Capacity Calculations**
   Strengths of concrete and rebars are first determined (see Rating Manual Section 6.6.2.3):

$$f_c' = \frac{f_c}{0.4} \quad \text{thus} \quad f_c' = 2500 \text{ psi}$$

and $f_s = 18{,}000$ psi and thus $f_y = 33{,}000$ psi.
   a. *Moment capacity at midspan*:

Total area of the steel (note these bars are 1¼ square bars)	$= (8)\,(1.25)\,(1.25) = 12.5$ in.^{2}
Centroid of the rebars from top deck	$= 42 + 8 - 3.75 = 46.250$ in.
Effective width of the deck b_{eff}	$=$ minimum of $12t_s + b_w = 112$ in.
Span/4	$= 150$ in.
Spacing	$= 80$ in. (Controls)

$$\text{Uniform stress block depth} = a = \frac{A_s f_y}{0.85 f_c' b_{eff}} = 2.426 \text{ in.} < t_s = 8 \text{ in.}$$

$$M_u = \phi A_s f_y \left(d - \frac{a}{2} \right) = 0.9 \times 12.5 \times 33 \left(46.25 - \frac{2.426}{2} \right) \times \left(\frac{1}{12} \right) = 1393.3 \text{ kips-ft } (1.88 \text{ MN-m})$$

   b. *Shear capacity at support*:
   According to AASHTO specification, shear at a distance d (50 in.) from the support needs to be designed. Thus, the girder is rated at a distance d from the support. From the girder details, it is estimated that ½ in. ϕ stirrups were placed at a spacing of 12 in. and two 1¼ square bars were bent up. The effects of these bent-up bars are ignored in the shear capacity calculations. Shear capacity due to concrete section:

$$V_c = 2\sqrt{f_c'}\, b_w d = 2\sqrt{2500} \times 16 \times 46.25 \times \left(\frac{1}{1000} \right) = 74 \text{ kips} \ (329 \text{ kN})$$

   Shear capacity due to shear reinforcement:

$$V_s = 2 A_v \frac{F_y d_s}{S} = 2 \times 0.20 \times \frac{33 \times 46.25}{12} = 50.88 \text{ kips } (226 \text{ kN})$$

   Total shear capacity:

$$V_u = \phi\,(V_s + V_c) = 0.85\,(74.0 + 50.88) = 106.2 \text{ kips } (472 \text{ kN})$$

4. **Rating Calculations**

$$\text{Rating factor} = \frac{\phi\, R_n - \gamma_D D}{\gamma_L \beta_L L(1+I)}$$

According to Rating Manual, γ_D is 1.3, γ_L is 1.3, and β_L is 1.67 and 1.0 for inventory and operating factors, respectively. By substituting these values and appropriate load effect values, the moment and shear rating could be estimated. The calculations and results are given in Table 5.1.

TABLE 5.1 Rating Calculations of Simply Supported T-Beam Concrete Bridge Example

Location	Description	Inventory Rating	Operating Rating
Midspan	Moment	$\dfrac{1393.3 - 1.3 \times 484.6}{1.3 \times 1.67 \times 425.7} = 0.825$	$\dfrac{1393.3 - 1.3 \times 484.6}{1.3 \times 425.7} = 1.38$
At support	Shear	$\dfrac{106.2 - 1.3 \times 32.31}{1.3 \times 1.67 \times 40.46} = 0.731$	$\dfrac{106.2 - 1.3 \times 32.31}{1.3 \times 40.46} = 1.22$

5. **Summary**

Critical rating of the interior girder will then be 0.731 at inventory level and 1.22 at operating rating level for HS20 vehicle.

5.6.3 Two-Span Continuous Steel Girder Bridge

Given

Typical section of a two-span continuous steel girder bridge, which was built in 1967, is shown in Figure 5.3a. Steel girder profile is given in Figure 5.3b. The general plan states that $f_s = 20,000$ psi (137.9 MPa) and $f_c = 1200$ psi (8.28 MPa). Assume that (a) each barrier rail weighs 250 lb/ft (3.6 N/mm); (b) girders were not temporarily supported during the concrete pour; (c) girder is composite for live loads; (d) girder is braced every 15 ft and the weight of bracing per girder is 330 lb.

Requirement

Determine the rating factors of interior girders using ASD method.

Solution

1. **Dead-Load Calculations**

   Deck weight (tributary area approach) = (6.625/12) (6.625) (0.15)   = 0.549 kips/ft

   Average uniform self-weight for the analysis = 1431 kips/90 ft   = 0.159 kips/ft

   Average diaphragm load (uniformly distributed) = (0.33) (4/90)   = 0.015 kips/ft

   Thus, total uniform dead load on the girder   = 0.723 kips/ft (10.5 N/mm)

   Barrier rail load (equally distributed among all girders) = (2)(250)/14 = 0.0358 kips/ft

   Thus, total additional dead load on the girder   = 0.0358 kips/ft (0.56 N/mm)

2. **Live-Load Calculations**

   Number of wheels per girder (for two or more lanes) = $S/5.5$ = 6.625/5.5 = 1.206

   Analysis Results: Analysis is done using two-dimensional program and the moments and shears at critical locations are listed in the Table 5.2. Section properties at 0.4th and 1.0th points are estimated and the results are given in Table 5.3.

3. **Allowable Stress Calculations**

   Strengths of concrete and rebars are first determined (see Rating Manual Section 6.6.2.3):

   $$f_c' = \frac{f_c}{0.4}, \text{ thus } f_c' = 3000 \text{ psi}$$

   and $f_s = 20,000$ psi and thus $F_y = 36,000$ psi.

   a. *Compression and tensile stresses at 0.4th point:*

   Note that the section is fully braced at this location.

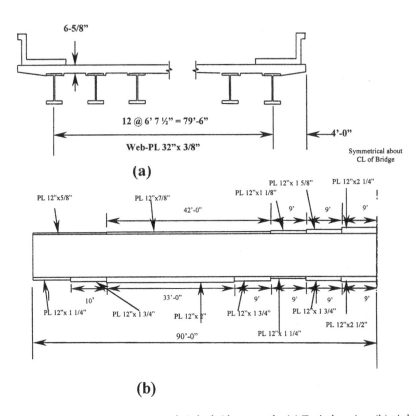

FIGURE 5.3 Details of two-span continuous steel girder bridge example. (a) Typical section; (b) girder elevation.

TABLE 5.2 Load Demands at 0.4th and 1.0th Point for Steel Girder Bridge Example

Description	0.4th Point (36.1 ft)	At Support (90 ft)
Dead-load moments in kip-ft	410.0	−732.0
Dead-load shear in kips	−1.7	−40.7
Additional dead-load moment in kip-ft	22.0	−39.0
Additional dead-load shear in kips	−0.1	−2.2
HS20 max. positive moment in kip-ft	807.0	0.0
HS20 max. negative moment in kip-ft	−177.0	−714.0
HS20 max. positive shear force in kips	21.8	0.0
HS20 max. negative shear force in kips	−19.3	−49.9

TABLE 5.3 Section Properties of Girder Sections for Steel Girder Bridge Example

	I_{gg} (in.^{4})	Y_b (in.)	Y_t (in.)	S_{xb} (in.^{3})	S_{xt} (in.^{3})
Section at 0.4th Point					
For dead loads	9,613.9	12.94	21.94	743.11	438.24
For additonal dead loads	17,406.5	19.69	15.19	884.18	1,146.10
For live loads	24,782.7	26.02	8.86	952.59	2,797.50
Section at 1.0th Point					
For dead loads	17,852.1	17.70	19.05	1,008.50	937.20
For additional dead loads	17,852.1	17.70	19.05	1,008.50	937.20
For live loads	17,852.1	17.70	19.05	1,008.50	937.20

     i.   Allowable compressive stress at inventory level = 0.55 F_y = 20 ksi (137.9 MPa)

     ii.   Allowable compressive stress at operating level = 0.75 F_y = 27 ksi (186.2 MPa)

     iii. Allowable tensile stress at inventory level = 0.55 F_y = 20 ksi (137.9 MPa)

  b. *Compression and tensile stresses at 1.0th point*:

     i.   Allowable tensile stress at inventory level = 0.55 F_y = 20 ksi (137.9 MPa)

     ii.   Allowable compressive stress at inventory level: Girder is braced 15 ft away from the support and thus L_b = 15 × 12 = 180 in. It can be shown that S_{xc} = 1008.3 in³; d = 36.75 in.; J = 108.63 in.⁴; I_{yc} = 360 in.⁴

           Then allowable stress at inventory level (Table 6.6.2.1-1 of Rating Manual):

$$F_b = \frac{91 \times 10^6 C_b}{1.82 \times S_{xc}} \left(\frac{I_{yc}}{L_b}\right) \sqrt{0.772\left(\frac{J}{I_{yc}}\right) + 9.87\left(\frac{d}{L_b}\right)^2}$$

$$= \frac{91 \times 10^6 (1.00)}{1.82 \times 1008.3} \left(\frac{360}{180}\right) \sqrt{0.772\left(\frac{108.63}{360}\right) + 9.87\left(\frac{36.75}{180}\right)^2} \left(\frac{1}{1000}\right)$$

$$= 79.5 > 0.55 F_y = 20 \text{ ksi (Note that } C_b \text{ is conservatively assumed as 1.0.)}$$

           Thus, F_b = 20 ksi (137.9 MPa)

     iii. Allowable compressive stress at operating level: The allowable stress at operating level is given: (Table 6.6.2.1-2, Rating Manual)

$$F_b = \frac{91 \times 10^6 C_b}{1.34 \times S_{sc}} \left(\frac{I_{yc}}{L_b}\right) \sqrt{0.772\left(\frac{J}{I_{yc}}\right) + 9.87\left(\frac{d}{L_b}\right)^2}$$

$$= 108.0 > 0.75 \ F_y = 27 \text{ ksi}$$

    Thus, F_b = 27 ksi (186.2 MPa)

  c. *Allowable inventory shear stresses at 0.4th and 1.0th point*:

$$D/t_w = 32/0.375 = 85.33$$

Girder is unstiffened and thus k = 5;

$$\frac{6000\sqrt{k}}{\sqrt{F_y}} = 70.71 < D/t_w < \frac{7500\sqrt{k}}{\sqrt{F_y}} = 88.3$$

Thus,

$$C = \frac{6000\sqrt{k}}{\left(\dfrac{D}{t_w}\right)\sqrt{F_y}} = 0.828$$

$$F_v = \frac{F_y}{3}\left(C + \frac{0.87(1-C)}{\sqrt{1+\left(\dfrac{d_o}{D}\right)^2}}\right) = 11.76 \text{ ksi (81.1 MPa)}$$

TABLE 5.4   Estimated Stress Demands for Steel Girder Bridge Example

Load Description	At 0.4th Point	At 1.0th Point	Fiber Location
DL moment	−6.62	8.71	At bottom fiber
ADL moment	−0.30	0.463	At bottom fiber
LL + *I* moment	−10.16	8.49	At bottom fiber
DL moment	11.23	−9.37	At top fiber
ADL moment	0.23	−0.49	At top fiber
LL + *I* moment	3.46	−9.14	At top fiber
DL shear	0.129	2.95	Shear stress
ADL shear	0.007	0.15	Shear stress
LL + *I* shear	1.667	3.62	Shear stress

d. *Allowable operating shear stresses at 0.4th and 1.0th point:*

$$F_v = 0.45 F_y \left(C + \frac{0.87(1-C)}{\sqrt{1+\left(\dfrac{d_o}{D}\right)^2}} \right) = 15.88 \text{ ksi } (109.5 \text{ MPa})$$

4. **Load Stress Calculations**

   Bending stress calculations are made using appropriate section modulus and moments. Results are reported in Table 5.4. The sign convention used in Table 5.4 is as follows: compressive stress is positive and tensile stress is negative. Also, estimated shear stresses are given in Table 5.4.

5. **Rating Calculations**

   The rating factor in ASD approach is given by

$$\frac{R-D}{L(1+I)}$$

   and the rating calculations are made and given in Table 5.5.

6. **Summary**

   The critical rating factor of the girder is controlled by tensile stress on the top fiber at the 1.0th point. The critical inventory and operating rating factors are 1.11 and 1.87, respectively.

TABLE 5.5   Rating Calculations Using ASD Method for Steel Girder Bridge Example

Location	Description	Inventory Rating	Operating Rating
0.4th point	Shear	$\dfrac{11.76-(0.129+0.007)}{1.667}=6.97$	$\dfrac{15.88-(0.129+0.007)}{1.667}=9.44$
	Stress at top fiber	$\dfrac{20-(11.23+0.23)}{3.46}=2.46$	$\dfrac{27-(11.23+0.23)}{3.46}=4.49$
	Stress at bottom fiber	$\dfrac{20-(6.62+0.30)}{10.16}=1.28$	$\dfrac{27-(6.62+0.30)}{10.16}=1.97$
1.0th point	Shear	$\dfrac{11.76-(2.95+0.15)}{3.62}=2.39$	$\dfrac{15.88-(2.95+0.15)}{3.62}=3.53$
	Stress at top fiber	$\dfrac{20-(9.37+0.50)}{9.14}=1.11$	$\dfrac{27-(9.37+0.50)}{9.14}=1.87$
	Stress at bottom fiber	$\dfrac{20-(8.71+0.463)}{8.49}=1.28$	$\dfrac{27-(8.71+0.463)}{8.49}=2.10$

5.6.4 Two-Span Continuous Prestressed, Precast Concrete Box Beam Bridge

Given

Typical section and elevation of three continuous-span precast, prestressed box-girder bridge is shown in Figure 5.4. The span length of each span is 120 ft (36.5 m), 133 ft (40.6 m), and 121 ft (36.9 m). Total width of the bridge is 82 ft (25 m) and a number of precast, prestressed box girders are placed at a spacing of 10 ft (3.1 m). The cross section of the box beam and the tendon profile of the girder are shown in Figure 5.4. Each barrier rail weighs 1268 lb/ft (18.5 N/mm). Information gathered from the plans is (a) f'_c of the girder and slab is 5500 and 3500 psi, respectively; (b) working force (total force remaining after losses including creep) = 2020 kips; (c) x at midspan = 9 in. Assume that (1) the bridge was made continuous for live loading; (2) no temporary supports were used during the erection of the precast box beams; (3) properties of the precast box are area = 1375 in²; moment of inertia = 30.84 ft⁴; Y_t = 28.58 in.; Y_b = 34.4 in.; (4) F_y of reinforcing steel is 40 ksi.

Requirement

Rate the interior girder of Span 2 for HS20 vehicle.

Solution

1. **Dead-Load Calculations**

Self-weight of the box beam = (1375/144) (0.15)   = 1.43 kips/ft

Weight of slab (tributary area approach) = (6.75/12) (10) (0.15) = 0.85 kips/ft

Total dead weight on the box beam   = 2.28 kips/ft (33.2 N/mm)

Contribution of barrier rail on box beam = 2 (1.268/8)   = 0.318 kips/ft

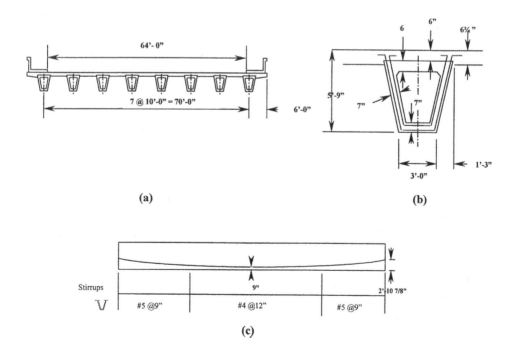

(a)   (b)

(c)

FIGURE 5.4 Details of two-span continuous prestressed box beam bridge example. (a) Typical section; (b) beam section details; (c) prestressing tendon profile.

Thus, total additional dead load on the box beam   = 0.318 kips/ft (4.6 N/mm)

Girder is simply supported for dead loads;

Thus maximum dead load moment = (2.28) (133²/8)   = 4926 kips/ft (6.68 MN/m)

2. **Live-Load Calculations**

According to Article 3.28 of Design Specifications, distribution factor (DF) for interior spread box beam is given by

$$DF = \left(\frac{2\,N_L}{N_B}\right) + k\left(\frac{S}{L}\right)$$

where N_L = number of traffic lanes = 64/12 = 5 (no fractions); N_B = Number of beams = 8; S = girder spacing = 8 ft; L = span length = 133 ft; W = roadway width = 64 ft

$$k = 0.07\ W - N_L\ (0.10\ N_L - 0.26) - 0.2\ N_B - 0.12 = 1.56$$

Thus,

$$DF = \left(\frac{2\times 5}{8}\right) + 1.56\left(\frac{10}{133}\right) = 1.37\ \text{wheels}$$

3. **Demands on the Girder**

Load demands are estimated using a two-dimensional analysis, and a summary is given in Table 5.6.

4. **Section Property Calculations**

In order to estimate the stresses on the prestress box beam, the section properties for composite girder need to be estimated. Calculations of the composite girder properties are done separately and the final results are listed here in Table 5.7.

5. **Stress Calculations**

Stresses at different fiber locations are calculated using

TABLE 5.6   Load Demands for Prestressed Precast Box Beam Bridge Example

Description	0.5L	At Bent 2	At Bent 3
Dead load moment (kip/ft)	4224	0	0
Additional dead-load moment (kip/ft)	194	−506	−513
HS20 moment with impact (kip/ft)	1142	−1313	−1322
Dead load shear (kips)	0.0	153.6	−153.6
Additional dead-load shear (kips)	0.0	21.1	−21.2
HS20 positive shear (moment)[a] (kips)	24.8 [1104]	61.1 [−974]	7.1 [127]
HS20 negative shear (moment)[a] (kips)	−24.8 [1104]	−7.1 [131]	−61.2 [−980]

   [a] Values within brackets indicate the moment corresponds to the reported shear.

TABLE 5.7   Section Properties for Prestressed, Precast Box Beam Bridge Example

Description	Area (in.²)	Moment of Inertia (ft⁴)	Y Bottom of Girder (in.)	Y Top of Girder (in.)	Y Top of Slab (in.)
For dead loads	1375	30.84	34.42	28.58	NA
For additional dead loads	1578	39.22	38.55	24.45	30.45
For live loads	1984	50.75	44.23	18.77	24.77

TABLE 5.8   Stresses at Midspan for Prestressed, Precast Box Beam Bridge Example

Location = Midspan	Stresses in the Box Beam (psi)			
Load Description	At Top Concrete Fiber	At Bottom Concrete Fiber	At Centroid of Composite Box Beam Concrete Fiber	At Prestress Tendon
Dead load (self + slab)	2265	−2728	777	20.15
Prestress P_{eff} = 2020 kips e = 25.42 in.	−1615	3443	−108	147.1
Additional dead (barrier)	70	−110	16	0.845
Live load	244	−575	0	4.59
Live-load moment for shear	236	−556	0	4.43

TABLE 5.9   Stresses at Bent 2 for Prestressed, Precast Box Beam Bridge Example

Location = Bent 2	Stresses in the Box Beam (psi)				
Load Description	At Top Concrete Fiber	At Bottom Concrete Fiber	At Centroid of Composite Box Beam Concrete Fiber	At Top of Slab Fiber	At Prestress Tendon
Dead load (self + slab)	0	0	0	0	0
Prestress P_{eff} = 2020 kips e = 12 in.	680	680	680	0	167.5
Additional dead (barrier)	−183	288	−4	−228	−0.3
Live load	−281	662	0	−371	−1.47
Live-load moment for positive shear	−208	491	0	−274	−1.08

$$\left(\frac{P}{A}\right) + \left(\frac{Mc}{I}\right)$$

The summary of the results at midspan and at Bent 2 locations is given in Tables 5.8 and Table 5.9, respectively.

6. **Capacity Calculations**
   a. *Moment capacity at midspan*:
      The actual area of steel could only be obtained from the shop plans. Since the shop plans are not readily available, the following approach is used. Assume the total loss including the creep loss = 35 ksi (241.3 MPa).

Thus, the area of prestressing steel = $\dfrac{\text{Working force}}{0.75 \times 270 - 35} = \dfrac{2020}{167.5} = 12.06 \text{ in.}^2$ (7781 mm²)

b_{eff} = 120 in.; t_s = 6.75 in.; d_p = (5.75)(12) − 9 in. = 60 in.; b_w = 14 in.

$$\rho^* = \frac{A_s^*}{bd} = \frac{12.06}{120 \times 60} = 0.001675$$

$$f_{su}^* = f_s \left(1 - \frac{0.5\rho^* f_s'}{f_c}\right) = 270\left(1 - \frac{0.5 \times 0.001675 \times 270}{5.5}\right) = 258.9 \text{ ksi } (1785 \text{ MPa})$$

Neutral axis location = $1.4d\rho^* \dfrac{f_{su}^*}{f_c'} = 1.4 \times 60 \times 0.001675 \times \dfrac{258.9}{5.5} = 6.62 \text{ in.} < t_s = 6.75 \text{ in.}$

Since the neutral axis falls within the slab, this girder can be treated as a rectangular section for moment capacity calculations.

$$R = \phi\, M_n = \phi\, A_s^* f_{su}^* d\left(1 - 0.6\rho \frac{f_{su}^*}{f_c'}\right) \quad \text{and} \quad \phi = 1.00$$

$$= 14873.1 \text{ kips/ft } (20.17 \text{ MN/m})$$

b. *Moment capacity at the face of the support:*
15 #11 bars are used at top of the bent; thus, the total area of steel = (15)(1.56) = 23.4 in.² Depth of the reinforcing steel from the top of compression fiber = 69 − 1.5 − 1.41/2 = 66.795 in. (1696.6 mm). F_y = 60 ksi. Resistance reduction factor ϕ = 0.90. Then, the moment capacity

$$\phi\, M_n = 6547.2 \text{ kip/ft } (8.88 \text{ MN/m}) \text{ (based on T section)}$$

c. *Shear capacity at midspan:*
Design Specification's Section 9.20 addresses the shear capacity of a section. Shear capacity depends on the cracking moment of the section. When the live load causes tension at bottom fiber, cracking moment is to be calculated based on the bottom fiber stress. On the other hand, when the live load causes tension at the top fiber of the beam, cracking moment is to be calculated based on the top fiber stress.

At midspan location, the moment reported with the maximum live-load shear is positive. Positive moments will induce tension at the bottom fiber and thus cracking moment is to be based on the stress at bottom fiber.

f_c' = 5500 psi and from Table 5.10; f_{pe} at midspan bottom fiber = 3443 psi

f_d at bottom fiber = −2728 − 110 = −2838 psi; f_{pc} at centroid = 777 − 108 + 16 = 685 psi

$$M_{cr} = \frac{I}{Y_t}\left(6\sqrt{f_c'} + f_{pe} - f_d\right) = \frac{50.75 \times 12^4}{44.23}\left(6\sqrt{5500} + 3443 - 2838\right)\left(\frac{1}{12,000}\right) = 2081 \text{ kips/ft}$$

Factored total moment:

$$M_{max} = 1.3\, M_D + (1.3)(1.67)\, M_{LL+I}$$

$$= 1.3\,(4224 + 194) + 2.167\,(1104) = 8136 \text{ kips-ft}$$

TABLE 5.10 Rating Calculations Prestressed, Precast Box Beam Bridge Example

Location	Description	Inventory Rating	Operating Rating
Midspan	Maximum moment	$\dfrac{14873.1 - 1.3 \times (4224 + 194)}{1.3 \times 1.67 \times 1142} = 3.69$	$\dfrac{14873.1 - 1.3 \times (4224 + 194)}{1.3 \times 1142} = 6.16$
	Maximum shear	$\dfrac{179 - 1.3 \times (0 + 0)}{1.3 \times 1.67 \times 24.8} = 3.33$	$\dfrac{179 - 1.3 \times (0 + 0)}{1.3 \times 24.8} = 5.56$
Bent 2	Maximum moment	$\dfrac{6544.2 - 1.3 \times (0 + 506)}{1.3 \times 1.67 \times 1313} = 2.06$	$\dfrac{6544.2 - 1.3 \times (0 + 506)}{1.3 \times 1313} = 3.45$
	Maximum shear	$\dfrac{766 - 1.3 \times (153.6 + 21.1)}{1.3 \times 1.67 \times 61.1} = 4.07$	$\dfrac{766 - 1.3 \times (153.6 + 21.1)}{1.3 \times 61.1} = 6.80$

Factored total shear:

$$V_i = 1.3 \ (0 + 0) + 2.167 \ (24.8) = 53.7 \text{ kips}$$

$$V_d = 0 \text{ kips;} \quad b_w = 14 \text{ in.;} \quad d = 60 \text{ in.;} \quad f_{pc} = 685 \text{ psi}$$

$$V_{ci} = 0.6\sqrt{f_c'}b_w d + V_d + \frac{V_i M_{cr}}{M_{max}} = 0.6\sqrt{5500}\times 14\times 60\times\left(\frac{1}{1000}\right)+0+\frac{53.7\times 2081}{8136}$$

$$= 51.2 \text{ kips (227.7 kN) (Controls — since smaller than } V_{cw})$$

$$V_{cw} = \left(3.5\sqrt{f_c'}+0.3f_{pc}\right)b_w d + V_p = \left(3.5\sqrt{5500}+0.3\times 685\right)14\times 60\times\left(\frac{1}{1000}\right)+0$$

$$= 390 \text{ kips (1734 kN)}$$

$$V_c = 51.2 \text{ kips (227.7 kN) (smaller of } V_{ci} \text{ and } V_{cw})$$

$$V_s = 2 \ A_v \frac{F_y d_s}{S} = 4\times 0.20\times\frac{40\times 60}{12} = 160 \text{ kips (711.7 kN)}$$

Shear capacity at midspan:

$$V_u = f \ (V_c + V_s) = 0.85 \ (51.2 + 160) = 179 \text{ kips (796.1 kN)}$$

d. *Shear capacity at the face of support at Bent 2:*
   Negative shear reported at this location is so small and thus rating will not be controlled by the negative shear at Bent 2. Moment reported with the positive shear is negative, and thus, the following calculations are based on the stress at top fiber. From Table 5.11, f_d at top of slab fiber = –228 psi, and f_{pe} at support top of slab fiber (slab poured after prestressing) = 0 psi.

$$M_{cr} = \frac{I}{Y_t}\left(6\sqrt{f_c'}+f_{pe}-f_d\right) = \frac{50.75\times 12^4}{44.23}\left(6\sqrt{3500}+0-228\right) = 252 \text{ kips/ft}$$

$$V_d = 153.6 + 21.1 = 174.7 \text{ kips;} \quad b_w = 14 \text{ in.;} \quad d = 69 - 1.5 - 1.41/2 = 66.795 \text{ in.;} \quad f_{pc} = 676 \text{ psi}$$

Factored total moment:

$$M_{max} = 1.3 \ M_D + (1.3)(1.67) \ M_{LL+I}$$

$$= 1.3(0 + -506) + 2.167(-974) = -2769 \text{ kips-ft}$$

Factored total shear:

$$V_i = 1.3 \times (153.6 + 21.1) + 2.167 \times (61.1) = 360 \text{ kips}$$

$$V_{ci} = 0.6\sqrt{5500}\times 14\times 66.795\left(\frac{1}{1000}\right)+0+\frac{360\times 251.7}{2769} = 74.3 \text{ kips}$$

$$V_{cw} = \left(3.5\sqrt{f_c'}+0.3f_{pc}\right)b_w d + V_p = \left(3.5\sqrt{5500}+0.3\times 676\right)14\times 66.795\left(\frac{1}{1000}\right)+0 = 432 \text{ kips}$$

$$V_c = 74.3 \text{ kips (330.4 kN)} \quad (\text{smaller of the } V_{cw} \text{ and } V_{ci})$$

$$V_s = 2 \ A_v \frac{F_y d_y}{S} = 4\times 0.31\times\frac{60\times 66.695}{6} = 827 \text{ kips (3678.5 kN)}$$

Shear capacity at Bent 2:

$$V_u = \phi\,(V_c + V_s) = 0.85\,(74.3 + 827) = 766 \text{ kips } (3408 \text{ kN})$$

7. **Rating Calculations**

   As discussed previously, the rating calculations for load factor method need to be done using strength and serviceability limit states. Serviceability level rating need not be done at the operating level.

   a. *Rating calculations based on serviceability limit state*:

      Serviceability conditions are listed in AASHTO Design Specification Sections 9.15.1 and 9.15.2.2. These conditions are duplicated in the Rating Manual.

      i.   Using the compressive stress under all load combination:

         The general expression will be $RF_{INV\text{-}COMALL} = \dfrac{0.6f_c' - f_d - f_p + f_s}{f_l}$

         At midspan $RF_{INV\text{-}COMALL} = \dfrac{0.6 \times 5500 - (2265 + 70) - (-1615) + 0}{244} = 10.57$

         At Bent 2 support $RF_{INV\text{-}COMALL} = \dfrac{0.6 \times 5500 - (0 + 288) - 680 + 0}{662} = 3.52$

      ii. Using the compressive stress of live load, half the prestressing and permanent dead load:

         The general expression will be $RF_{INV\text{-}COMLIVE} = \dfrac{0.4f_c' - f_d - 0.5f_p + 0.5f_s}{f_l}$

         At midspan $RF_{INV\text{-}COMLIVE} = \dfrac{0.4 \times 5500 - (2265 + 70) - 0.5(-1615) + 0.5(0)}{244} = 2.76$

         At Bent 2 support $RF_{INV\text{-}COMLIVE} = \dfrac{0.4 \times 5500 - (0 + 288) - 0.5(680) + 0.5(0)}{662} = 2.37$

      iii. Using the allowable tension in concrete:

         The general expression will be $RF_{INV\text{-}CONTEN} = \dfrac{6\sqrt{f_c'} - f_d - f_p - f_s}{f_l}$

         At midspan $RF_{INV\text{-}CONTEN} = \dfrac{6\sqrt{5500} - (2728 + 110) - (-3443) - 0}{575} = 1.826$

         At Bent 2 support $RF_{INV\text{-}CONTEN} = \dfrac{6\sqrt{5500} - (0 + 183) - (-680) - 0}{281} = 3.352$

      iv. Using the allowable prestressing steel tension at service level:

         The general expression will be $RF_{INV\text{-}PRETEN} = \dfrac{0.8f_y^* - f_d - f_p - f_s}{f_l}$

         At midspan $RF_{INV\text{-}PRETEN} = \dfrac{0.8 \times 270 - 20.99 - (147.1) - 0}{4.59} = 10.43$

At Bent 2 support $RF_{INV\text{-}PRETEN} = \dfrac{0.8\times270-(-3.08)-167.5-0}{1.468} = 30.94$

b. Rating calculations based on strength limit state:

The general expression for Rating factor $= \dfrac{\phi\,R_n-\gamma_D D}{\gamma_L\beta_L L(1+I)}$

According to AASHTO Rating Manual, γ_D is 1.3, γ_L is 1.3, and β_L is 1.67 and 1.0 for inventory and operating factor, respectively. Rating calculations are made and given in Table 5.10.

8. **Summary**

The critical inventory rating of the interior girder is controlled by the tensile stress on concrete at midspan location. The critical operating rating of the girder is controlled by moment at Bent 2 location.

5.6.5 Bridges without Plans

There are some old bridges in service without plans. Establishing safe live-load-carrying capacity is essential to have a complete bridge document. When an inspector comes across a bridge without plans, sufficient field physical dimensions of each member and overall bridge geometry should be taken and recorded. In addition, information such as design year, design vehicle, designer, live-load history, and field condition of the bridge needs to be collected and recorded. This information will be very helpful to determine the safe live-load-carrying capacity. Also, bridge inspectors need to establish the material strength either using the design year or coupon testing.

Design vehicle information could be established based on the designer (state or local agency) and the design year. For example, all state bridges have been designed using the HS20 vehicle since 1944 and all local agency bridges have been designed using the H15 vehicle since 1950.

In steel girder bridges, section properties of the members could be determined based on the field dimensions. During the estimation of the moment capacity, it is recommended to assume that the steel girders are noncomposite with the slab unless substantial evidence is gathered to prove otherwise.

In concrete girder bridges, field dimensions help to estimate the dead loads on the girders. Since the area of reinforcing steels is not known or is difficult to establish, determining the safe live load poses challenges to bridge owners. The live-load history and field condition of a bridge could be used to establish the safe load capacity of the bridge. For example, if a particular bridge has been carrying several heavy vehicles for years without damaging the bridge, this bridge could be left open for all legal vehicles.

5.7 Posting of Bridges

Bridge inspection and the strength evaluation process are two integral parts of bridge posting. The purpose of bridge inspection is to obtain the information that is necessary to evaluate the bridge capacity and the adequacy of the bridge properly. When a bridge is found to have inadequate capacity for legal vehicles, engineers need to look at several alternatives prior to closing the bridge to the public. Some of the possible alternatives are imposing speed limits, reducing vehicular traffic, limiting or posting for vehicle weight, restricting the vehicles to certain lanes, recommending possible small repairs to alleviate the problem. In addition, when the evaluations show that the structure is marginally inadequate, frequent inspections to monitor the physical condition of the bridge and traffic flow may be recommended.

Standard evaluation methods described in the Section 5.5 may be overly conservative. When a more accurate answer is required, a more-detailed analysis such as three-dimensional analysis or physical load testing can be performed.

The weight and axle configuration of vehicles allowed to use highways without special permits is governed by the statutory law. Thus, the traffic live loads used for posting purposes should be representative of the actual vehicles using the bridge. The representative vehicles vary with each state in the United States. Several states use the three hypothetical legal vehicle configurations given in the Rating Manual [12]. Whereas a few states use their own specially developed legal truck configurations, AASHTO H or HS design trucks, or some combination of truck types. NBIS requires that posting of a bridge must be done when the operating rating for three hypothetical legal vehicles listed in the Rating Manual [12] is less than unity. Furthermore, the NBIS requirement allows the bridge owner to post a bridge for weight limits between inventory and operating level. Because of this flexible NBIS requirement, there is a considerable variation in posting practices among various state and local jurisdictions.

Although engineers may recommend one or a combination of the alternatives described above, it is the owner, not the engineer, who ultimately makes the decision. Many times, bridges are posted for reasons other than structural evaluation, such as posting at a lower weight level to limit vehicular or truck traffic, posting at a higher weight level when the owner believes a lower posting would not be prudent and is willing to accept a higher level of risk. Weight limit posting may cause inconvenience and hardship to the public. In order to reduce inconvenience to the public, the owner needs to look at the weight limit posting as a last resort. In addition, it is sometimes in the public interest to allow certain overweight vehicles such as firefighting equipment and snow removal equipment on a posted bridge. This is usually done through the use of special permits.

References

1. FHWA, Bridge Inspector's Training Manual 70, U.S. Department of Transportation, Washington, D.C., 1970.
2. AASHO, *Manual for Maintenance Inspection of Bridges* American Association of State Highway Officials, Washington, D.C., 1970.
3. FHWA, Recording and Coding Guide for the Structure Inventory and Appraisal of the Nation's Bridges, U.S. Department of Transportation, Washington, D.C., 1972.
4. FHWA, Bridge Inspector's Manual for Movable Bridges, (Supplement to Manual 70), U.S. Department of Transportation, Washington, D.C., 1970.
5. FHWA, Culvert Inspection Manual (Supplement to Manual 70), U.S. Department of Transportation, Washington, D.C., 1970.
6. FHWA, Inspection of Fracture Critical Bridge Members, U.S. Department of Transportation, Washington, D.C., 1970.
7. FHWA, Inspection of Fracture Critical Bridge Members, U.S. Department of Transportation, Washington, D.C., 1986.
8. FHWA, Recording and Coding Guide for the Structure Inventory and Appraisal of the Nation's Bridges, U.S. Department of Transportation, Washington, D.C., 1979.
9. FHWA, Recording and Coding Guide for the Structure Inventory and Appraisal of the Nation's Bridges, U.S. Department of Transportation, Washington, D.C., 1988.
10. FHWA, Bridge Inspector's Training Manual 90, U.S. Department of Transportation, Washington, D.C., 1991.
11. FHWA, Hydraulic Engineering Circular (HEC) No. 18, Evaluating Scour at Bridges, U.S. Department of Transportation, Washington, D.C., 1990.
12. AASHTO, *Manual for Condition Evaluation of Bridges 1994*, American Association of State Highway and Transportation Officials, Washington, D.C., 1994.
13. AASHTO, *Standard Specification for Highway Bridges*, 16th ed., American Association of State Highway and Transportation Officials, Washington, D.C., 1996.

6

Strengthening and Rehabilitation

F. Wayne Klaiber
Iowa State University

Terry J. Wipf
Iowa State University

6.1 Introduction

About one half of the approximately 600,000 highway bridges in the United States were built before 1940, and many have not been adequately maintained. Most of these bridges were designed for lower traffic volumes, smaller vehicles, slower speeds, and lighter loads than are common today. In addition, deterioration caused by environmental factors is a growing problem. According to the Federal Highway Administration (FHWA), almost 40% of the nation's bridges are classified as deficient and in need of rehabilitation or replacement. Many of these bridges are deficient because their load-carrying capacity is inadequate for today's traffic. Strengthening can often be used as a cost-effective alternative to replacement or posting.

The live-load capacity of various types of bridges can be increased by using different methods, such as (1) adding members, (2) adding supports, (3) reducing dead load, (4) providing continuity, (5) providing composite action, (6) applying external post-tensioning, (7) increasing member cross section, (8) modifying load paths, and (9) adding lateral supports or stiffeners. Some methods have been widely used, but others are new and have not been fully developed.

All strengthening procedures presented in this chapter apply to the superstructure of bridges. Although bridge span length is not a limiting factor in the various strengthening procedures presented, the majority of the techniques apply to short-span and medium-span bridges. Several of the strengthening techniques, however, are equally effective for long-span bridges. No information is included on the strengthening of existing foundations because such information is dependent on soil type and conditions, type of foundation, and forces involved.

The techniques used for strengthening, stiffening, and repairing bridges tend to be interrelated so that, for example, the stiffening of a structural member of a bridge will normally result in its being strengthened also. To minimize misinterpretation of the meaning of strengthening, stiffening, and repairing, the authors' definitions of these terms are given below. In addition to these terms, definitions of maintenance and rehabilitation, which are sometimes misused, are also given.

Maintenance: The technical aspect of the upkeep of the bridges; it is preventative in nature. Maintenance is the work required to keep a bridge in its present condition and to control potential future deterioration.

Rehabilitation: The process of restoring the bridge to its original service level.

Repair: The technical aspect of rehabilitation; action taken to correct damage or deterioration on a structure or element to restore it to its original condition.

Stiffening: Any technique that improves the in-service performance of an existing structure and thereby eliminates inadequacies in serviceability (such as excessive deflections, excessive cracking, or unacceptable vibrations).

Strengthening: The increase of the load-carrying capacity of an existing structure by providing the structure with a service level higher than the structure originally had (sometimes referred to as upgrading).

In recent years the FHWA and National Cooperative Highway Research Program (NCHRP) have sponsored several studies on bridge repair, rehabilitation, and retrofitting. Inasmuch as some of these procedures also increase the strength of a given bridge, the final reports on these investigations are excellent references. These references, plus the strengthening guidelines presented in this chapter, will provide information an engineer can use to resolve the majority of bridge strengthening problems. The FHWA and NCHRP final reports related to this investigation are references [1–13].

Four of these references [1, 2, 11, 12] are of specific interest in strengthening work. Although not discussed in this chapter, the live-load capacity of a given bridge can often be evaluated more accurately by using more-refined analysis procedures. If normal analytical methods indicate strengthening is required, frequently more-sophisticated analytical methods (such as finite-element analysis) may result in increased live-load capacities and thus eliminate the need to strengthen or significantly decrease the amount of strengthening required.

By load testing bridges, one frequently determines live-load capacities considerably larger than what one would determine using analytical procedures. Load testing of bridges makes it possible to take into account several contributions (such as end restraint in simple spans, structural contributions of guardrails, etc.) that cannot be included analytically. In the past few years, several states have started using load testing to establish live-load capacities of their bridges. An excellent reference on this procedure is the final report for NCHRP Project 12-28(13)A [14]. Most U.S. states have some type of bridge management system (BMS). To the authors' knowledge, very few states are using their BMS to make bridge strengthening decisions. At the present time, there are not sufficient baseline data (first cost, life cycle costs, cost of various strengthening procedures, etc.) to make strengthening/replacement decisions.

Examination of National Bridge Inventory (NBI) bridge records indicates that the bridge types with greatest potential for strengthening are steel stringer, timber stringer, and steel through-truss. If rehabilitation and strengthening cannot be used to extend their useful lives, many of these bridges will require replacement in the near future. Other bridge types for which there also is potential for strengthening are concrete slab, concrete T, concrete stringer, steel girder floor beam, and concrete deck arch. In this chapter, information is provided on the more commonly used strengthening procedures as well as a few of the new procedures that are currently being researched.

6.2 Lightweight Decks

6.2.1 Introduction

One of the more fundamental approaches to increase the live-load capacity of a bridge is to reduce its dead load. Significant reductions in dead load can be obtained by removing an existing heavier concrete deck and replacing it with a lighter-weight deck. In some cases, further reduction in dead load can be obtained by replacing the existing guardrail system with a lighter-weight guardrail. The concept of strengthening by dead-load reduction has been used primarily on steel structures, including the following types of bridges: steel stringer and multibeam, steel girder and floor beam, steel truss, steel arch, and steel suspension bridges; however, this technique could also be used on bridges constructed of other materials.

Lightweight deck replacement is a feasible strengthening technique for bridges with structurally inadequate, but sound, steel stringers or floor beams. If, however, the existing deck is not in need of replacement or extensive repair, lightweight deck replacement would not be economically feasible.

Lightweight deck replacement can be used conveniently in conjunction with other strengthening techniques. After an existing deck has been removed, structural members can readily be strengthened, added, or replaced. Composite action, which is possible with some lightweight deck types, can further increase the live-load carrying capacity of a deficient bridge.

6.2.2 Types

Steel grid deck is a lightweight flooring system manufactured by several firms. It consists of fabricated, steel grid panels that are field-welded or bolted to the bridge superstructure. The steel grids may be filled with concrete, partially filled with concrete, or left open (Figure 6.1).

Open-Grid Steel Decks

Open-grid steel decks are lightweight, typically weighing 15 to 25 psf (720 to 1200 Pa) for spans up to 5 ft (1.52 m). Heavier decks, capable of spanning up to 9 ft (2.74 m), are also available; the percent increase in live-load capacity is maximized with the use of an open-grid steel deck. Rapid installation is possible with the prefabricated panels of steel grid deck. Open-grid steel decks also have the advantage of allowing snow, water, and dirt to wash through the bridge deck, thus eliminating the need for special drainage systems.

A disadvantage of the open grids is that they leave the superstructure exposed to weather and corrosive chemicals. The deck must be designed so water and debris do not become trapped in the grids that rest on the stringers. Other problems associated with open-steel grid decks include weld failure and poor skid resistance. Weld failures between the primary bearing bars of the deck and the supporting structure have caused maintenance problems with some open-grid decks. The number of weld failures can be minimized if the deck is properly erected.

In an effort to improve skid resistance, most open-grid decks currently on the market have serrated or notched bars at the traffic surface. Small studs welded to the surface of the steel grids have also been used to improve skid resistance. While these features have improved skid resistance, they have not eliminated the problem entirely [12]. Open-grid decks are often not perceived favorably by the general public because of the poor riding quality and increased tire noise.

Concrete-Filled Steel Grid Decks

Concrete-filled steel grid decks weigh substantially more, but have several advantages over the open-grid steel decks, including increased strength, improved skid resistance, and better riding quality. The steel grids can be either half or completely filled with concrete. A 5-in. (130-mm) thick, half-filled steel grid weighs 46 to 51 psf (2.20 to 2.44 kPa), less than half the weight of a reinforced concrete deck of comparable strength. Typical weights for 5-in. (130-mm) thick steel grid decks, filled to full depth with concrete, range from 76 to 81 psf (3.64 to 3.88 kPa). Reduction in the deadweight resulting from concrete-filled

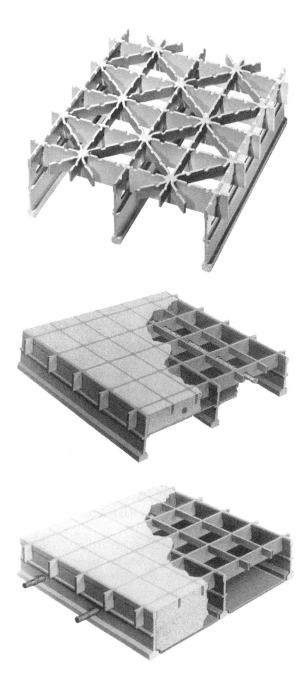

FIGURE 6.1 Steel-grid bridge deck. Top photo shows open steel grid deck; center photo shows half-filled steel grid deck; bottom photo shows filled steel grid deck. (*Source:* Klaiber, F.W. et al., NCHRP 293, Transportation Research Board, 1987. With permission.)

steel grid deck replacement alone only slightly improves the live-load capacity; however, the capacity can be further improved by providing composite action between the deck and stringers.

Steel grid panels that are filled or half-filled with concrete may either be precast prior to erection or filled with concrete after placement. With the precast system, only the grids that have been left open to allow field welding of the panels must be filled with concrete after installation. The precast system is generally used when erection time must be minimized.

A problem that has been associated with concrete-filled steel grid decks, addressed in a study by Timmer [15], is the phenomenon referred to as deck growth — the increase in length of the filled grid deck caused by the rusting of the steel I-bar webs. The increase in thickness of the webs due to rusting results in comprehensive stresses in the concrete fill. Timmer noted that in the early stages of deck growth, a point is reached when the compression of the concrete fill closes voids and capillaries in the concrete. Because of this action, the amount of moisture that reaches the resting surfaces is reduced and deck growth is often slowed down or even halted. If, however, the deck growth continues beyond this stage, it can lead to breakup of the concrete fill, damage to the steel grid deck, and possibly even damage to the bridge superstructure and substructure. Timmer's findings indicate that the condition of decks that had been covered with some type of wearing surface was superior to those that had been left unsurfaced. A wearing surface is also recommended to prevent wearing and eventual cupping of the concrete between the grids.

Exodermic Deck

Exodermic deck is a recently developed, prefabricated modular deck system that has been marketed by major steel grid deck manufacturers. The first application of Exodermic deck was in 1984 on the Driscoll Bridge located in New Jersey [16]. As shown in Figure 6.2, the bridge deck system consists of a thin upper layer, 3 in. (76 mm) minimum, of prefabricated concrete joined to a lower layer of steel grating. The deck weighs from 40 to 60 psf (1.92 to 2.87 kPa) and is capable of spanning up to 16 ft (4.88 m).

Exodermic decks have not exhibited the fatigue problems associated with open-grid decks or the growth problems associated with concrete-filled grid decks. As can be seen in Figure 6.2, there is no concrete fill and thus no grid corrosion forces. This fact, coupled with the location of the neutral axis, minimizes the stress at the top surface of the grid.

Exodermic deck and half-filled steel grid deck have the highest percent increase in live-load capacity among the lightweight deck types with a concrete surface. As a prefabricated modular deck system, Exodermic deck can be quickly installed. Because the panels are fabricated in a controlled environment, quality control is easier to maintain and panel fabrication is independent of the weather or season.

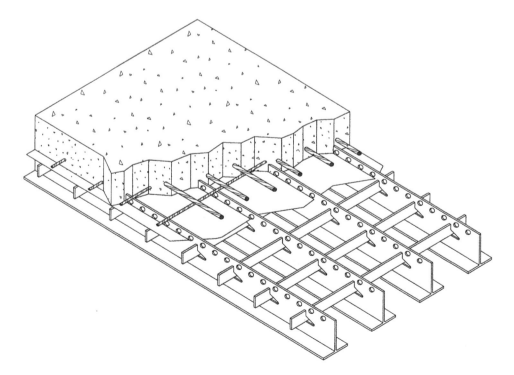

FIGURE 6.2 Exodermic deck system. (*Source:* Exodermic Bridge Deck Inc., Lakeville, CT, 1999. With permission.)

Laminated Timber Deck

Laminated timber decks consist of vertically laminated 2-in. (51-mm) (nominal) dimension lumber. The laminates are bonded together with a structural adhesive to form panels that are approximately 48 in. (1.22 m) wide. The panels are typically oriented transverse to the supporting structure of the bridge (Figure 6.3). In the field, adjacent panels are secured to each other with steel dowels or stiffener beams to allow for load transfer and to provide continuity between the panels.

A steel–wood composite deck for longitudinally oriented laminates has been developed by Bakht and Tharmabala [17]. Individual laminates are transversely post-tensioned in the manner developed by Csagoly and Taylor [18]. The use of shear connectors provides partial composite action between the deck and stringers. Because the deck is placed longitudinally, diaphragms mounted flush with the stringers may be required for support. Design of this type of timber deck is presented in References [19–21].

The laminated timber decks used for lightweight deck replacement typically range in depth from 3⅛ to 6¾ in. (79 to 171 mm) and from 10.4 to 22.5 psf (500 to 1075 Pa) in weight. A bituminous wearing surface is recommended.

Wood is a replenishable resource that offers several advantages: ease of fabrication and erection, high strength-to-weight ratio, and immunity to deicing chemicals. With the proper treatment, heavy timber members also have excellent thermal insulation and fire resistance [22]. The most common problem associated with wood as a structural material is its susceptibility to decay caused by living fungi, wood-

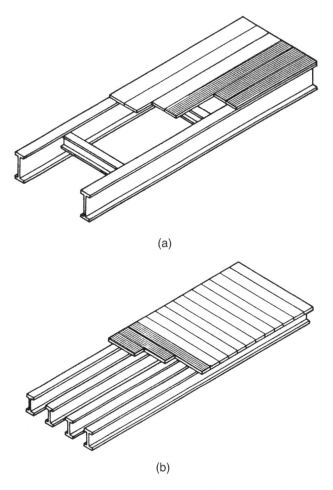

(a)

(b)

FIGURE 6.3 Laminated timber deck. (a) Longitudinal orientation; (b) transverse orientation. (*Source:* Klaiber, F.W. et al., NCHRP 293, Transportation Research Board, 1987. With permission.)

boring insects, and marine organisms. With the use of modern preservative pressure treatments, however, the expected service life of timber decks can be extended to 50 years or more.

Lightweight Concrete Deck

Structural lightweight concrete, concrete with a unit weight of 115 pcf (1840 kg/m³) or less, can be used to strengthen steel bridges that have normal-weight, noncomposite concrete decks. Special design considerations are necessary for lightweight concrete. Its modulus of elasticity and shear strength are less than that of normal-weight concrete, whereas its creep effects are greater [23]. The durability of lightweight concrete has been a problem in some applications.

Lightweight concrete for deck replacement can be either cast in place or installed in the form of precast panels. A cast-in-place lightweight concrete deck can easily be made to act compositely with the stringers. The main disadvantage of a cast-in-place concrete deck is the length of time required for concrete placement and curing.

Lightweight precast panels, fabricated with either mild steel reinforcement or transverse prestressing, have been used in deck replacement projects to help minimize erection time and resulting interruptions to traffic. Precast panels require careful installation to prevent water leakage and cracking at the panel joints. Composite action can be attained between the deck and the superstructure; however, some designers have chosen not to rely on composite action when designing a precast deck system.

Aluminum Orthotropic Plate Deck

Aluminum orthotropic deck is a structurally strong, lightweight deck weighing from 20 to 25 psf (958 to 1197 Pa). A proprietary aluminum orthotropic deck system that is currently being marketed is shown in Figure 6.4. The deck is fabricated from highly corrosion-resistant aluminum alloy plates and extrusions that are shop-coated with a durable, skid-resistant, polymer wearing surface. Panel attachments between the deck and stringer must not only resist the upward forces on the panels, but also allow for the differing thermal movements of the aluminum and steel superstructure. For design purposes, the manufacturer's recommended connection should not be considered to provide composite action.

The aluminum orthotropic plate is comparable in weight to the open-grid steel deck. The aluminum system, however, eliminates some of the disadvantages associated with open grids: poor ridability and acoustics, weld failures, and corrosion caused by through drainage. A wheel-load distribution factor has not been developed for the aluminum orthotropic plate deck at this time. Finite-element analysis has been used by the manufacturer to design the deck on a project-by-project basis.

Steel Orthotropic Plate Deck

Steel orthotropic plate decks are an alternative for lightweight deck replacements that generally have been designed on a case-by-case basis, without a high degree of standardization. The decks often serve several functions in addition to carrying and distributing vertical live loads and, therefore, a simple reinforced concrete vs. steel orthotropic deck weight comparison could be misleading.

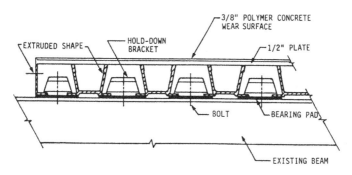

FIGURE 6.4 Aluminum orthotropic deck. (*Source:* Klaiber, F.W. et al., NCHRP 293, Transportation Research Board, 1987. With permission.)

Originally, steel orthotropic plate decks were developed to minimize steel use in 200- to 300-ft (61- to 91-m) span girder bridges. Then the decks were used in longer-span suspension and cable-stayed bridges where the deck weight is a significant part of the total superstructure design load. Although the steel orthotropic deck is applicable for spans as short as 80 to 120 ft (24.4 to 36.6 m), it is unlikely that there would be sufficient weight savings at those spans to make it economical to replace a reinforced concrete deck with a steel orthotropic plate deck. Orthotropic steel decks are heavier than aluminum orthotropic decks and usually have weights in the 45 to 130 psf (2.15 to 6.22 kPa) range.

6.2.3 Case Studies

Steel Grid Deck

The West Virginia Department of Highways was one of the first to develop a statewide bridge rehabilitation plan using open-grid steel deck [24]. By 1974, 25 bridges had been renovated to meet or exceed AASHTO requirements. Deteriorated concrete decks were replaced with lightweight, honeycombed steel grid decks fabricated from ASTM A588 steel. The new bridge floors are expected to have a 50-year life and to require minimal maintenance.

In 1981, the West Virginia Department of Highways increased the live-load limit on a 1794-ft (546.8-m) long bridge over the Ohio River from 3 tons (26.69 kN) to 13 tons (115.65 kN) by replacing the existing reinforced concrete deck with an open steel grid deck [25, 26]. The existing deck was removed and the new deck installed in sections allowing half of the bridge to be left open for use by workers, construction vehicles, and equipment, and, if needed, emergency vehicles.

The strengthening of the 250-ft (76.2-m) long Old York Road Bridge in New Jersey in the early 1980s combined deck replacement with the replacement of all of the main framing members and the modernization of the piers and abutments [27]. The existing deck was replaced with an ASTM A588 open-grid steel deck. The posted 10-ton (89-kN) load limit was increased to 36 tons (320 kN) and the bridge was widened from 18 ft (5.49 m) to 26 ft (7.92 m).

Exodermic Deck

The first installation of exodermic deck was in 1984 on the 4400-ft (1340-m) long Driscoll Bridge located in New Jersey [16]. The deck, weighing 53 psf (2.54 kPa), consisted of a 3-in. (76-mm) upper layer of prefabricated reinforced concrete joined to a lower layer of steel grating. Approximately 30,000 ft^2 (2790 m^2) of deck was replaced at this site.

Exodermic deck was also specified for the deck replacement on a four-span bridge which overpasses the New York State Thruway [28]. The bridge was closed to traffic during deck removal and replacement. Once the existing deck has been removed, it is estimated that approximately 7500 ft^2 (697 m^2) of exodermic deck will be installed in 3 working days.

Lightweight Concrete Deck

Lightweight concrete was used as early as 1922 for new bridge construction in the United States. Over the years, concrete made with good lightweight aggregate has generally performed satisfactorily; however, some problems related to the durability of the concrete have been experienced. The Louisiana Department of Transportation has experienced several deck failures on bridges built with lightweight concrete in the late 1950s and early 1960s. The deck failures have typically occurred on bridges with high traffic counts and have been characterized by sudden and unexpected collapse of sections of the deck.

Lightweight concrete decks can either be cast in place or factory precast. Examples of the use of lightweight concrete for deck replacement follow.

Cast-in-Place Concrete

New York state authorities used lightweight concrete to replace the deck on the north span of the Newburgh–Beacon Bridge [8, 29]. The existing deck was replaced with 6½ in. (165 mm) of cast-in-place lightweight concrete that was surfaced with a 1½ in. (38 mm) layer of latex modified concrete. Use of the lightweight concrete allowed the bridge to be widened from two to three lanes with minimal modifications to the substructure. A significant reduction in the cost of widening the northbound bridge was attributed to the reduction in dead load.

Precast Concrete Panels

Precast modular-deck construction has been used successfully since 1967 when a joint study, conducted by Purdue University and Indiana State Highway Commission, found precast, prestressed deck elements to be economically and structurally feasible for bridge deck replacement [30, 31].

Precast panels, made of lightweight concrete, 115 pcf (1840 kg/m³), were used to replace and widen the existing concrete deck on the Woodrow Wilson Bridge, located on Interstate 95 south of Washington, D.C. [32, 33]. The precast panels were transversely prestressed and longitudinally post-tensioned. Special sliding steel-bearing plates were used between the panels and the structural steel to prevent the introduction of unwanted stresses in the superstructure. The Maryland State Highway Commission required that all six lanes of traffic be maintained during the peak traffic hours of the morning and evening. Two-way traffic was maintained at night when the removal and replacement of the deck was accomplished.

Aluminum Orthotropic Plate Deck

The 104-year-old Smithfield Street Bridge in Pittsburgh, Pennsylvania, has undergone two lightweight deck replacements, both involving aluminum deck [34]. The first deck replacement occurred in 1933 when the original heavyweight deck was replaced with an aluminum deck and floor framing system. The aluminum deck was coated with a 1½-in. (38-mm) asphaltic cement wearing surface. The new deck, weighing 30 psf (1.44 kPa), eliminated 751 tons (6680 kN) of deadweight and increased the live-load capacity from 5 tons (44.5 kN) to 20 tons (178 kN).

Excessive corrosion of some of the deck panels and framing members necessitated the replacement of the aluminum deck on the Smithfield Street Bridge in 1967. At that time, a new aluminum orthotropic plate deck with a ⅜-in. (9.5-mm)-thick polymer concrete wearing surface was installed. This new deck weighed 15 psf (718 Pa) and resulted in an additional 108-ton (960-kN) reduction in deadweight. The panels were originally attached to the structure with anodized aluminum bolts, but the bolts were later replaced with galvanized steel bolts after loosening and fracturing of the aluminum bolts became a problem. The aluminum components of the deck have shown no significant corrosion; however, because of excessive wear, the wearing surface had to be replaced in the mid-1970s. The new wearing surface consisted of aluminum-expanded mesh filled with epoxy resin concrete. This wearing surface has also experienced excessive wear, and thus early replacement is anticipated.

Steel Orthotropic Plate Deck

Steel orthotropic plate decks were first conceived in the 1930s for movable bridges and were termed battledecks. Steel orthotropic decks were rapidly developed in the late 1940s in West Germany for replacement of bridges destroyed in World War II during a time when steel was in short supply, and replacement of bridge decks with steel orthotropic plate decks became a means for increasing the live-load capacity of medium- to long-span bridges in West Germany in the 1950s.

In 1956 Woeltinger and Bock [35] reported the rebuilding of a wrought iron, 536-ft (163-m) span bridge near Kiel. The two-hinged, deck arch bridge, which carried both rail and highway traffic, was widened and strengthened through rebuilding essentially all of the bridge except the arches and abutments. The replacement steel orthotropic deck removed approximately 190 tons of dead load from the bridge, improved the deck live-load capacity, and was constructed in such a way as to replace the original lateral wind bracing truss.

The live-load class of a bridge near Darmstadt was raised by means of a replacement steel orthotropic deck also in the mid-1970s [36]. The three-span, steel-through-truss bridge had been repaired and altered twice since World War II, but the deck had finally deteriorated to the point where it required replacement. The existing reinforced concrete deck was then replaced with a steel orthotropic plate deck, and the reduction in weight permitted the bridge to be reclassified for heavier truck loads.

6.3   Composite Action

6.3.1   Introduction

Modification of an existing stringer and deck system to a composite system is a common method of increasing the flexural strength of a bridge. The composite action of the stringer and deck not only reduces the live-load stresses but also reduces undesirable deflections and vibrations as a result of the increase in the flexural stiffness from the stringer and deck acting together. This procedure can also be used on bridges that only have partial composite action, because the shear connectors originally provided are inadequate to support today's live loads.

The composite action is provided through suitable shear connection between the stringers and the roadway deck. Although numerous devices have been used to provide the required horizontal shear resistance, the most common connection used today is the welded stud.

6.3.2   Applicability and Advantages

Inasmuch as the modifications required for providing composite action for continuous spans and simple spans are essentially the same, this section is written for simple spans. Composite action can effectively be developed between steel stringers and various deck materials, such as normal-weight reinforced concrete (precast or cast in place), lightweight reinforced concrete (precast or cast in place), laminated timber, and concrete-filled steel grids. These are the most common materials used in composite decks; however, there are some instances in which steel deck plates have been made composite with steel stringers. In the following paragraphs these four common deck materials will be discussed individually.

Because steel stringers are normally used for support of all the mentioned decks, they are the only type of superstructure reviewed. The condition of the deck determines how one can obtain composite action between the stringers and an existing concrete deck. If the deck is badly deteriorated, composite action is obtained by removing the existing deck, adding appropriate shear connectors to the stringers, and recasting the deck. This was done in Blue Island, Illinois, on the 1500-ft (457-m) long steel plate girder Burr Oak Avenue Viaduct [37].

If it is desired to reduce interruption of traffic, precast concrete panels are one of the better solutions. The panels are made composite by positioning holes formed in the precast concrete directly over the structural steel. Welded studs are then attached through the preformed holes. This procedure was used on an I-80 freeway overpass near Oakland, California [38]. As shown in Figure 6.5, panels 30 ft (9.1 m)

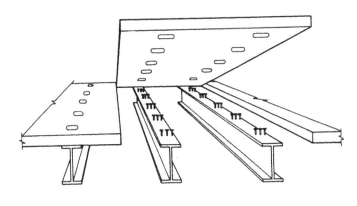

NOTE: SHEAR STUDS SHOWN ARE
ACTUALLY ADDED AFTER
PRECAST DECK IS POSITIONED.

FIGURE 6.5 Precast deck with holes. (*Source:* Klaiber, F.W. et al., NCHRP 293, Transportation Research Board, 1987. With permission.)

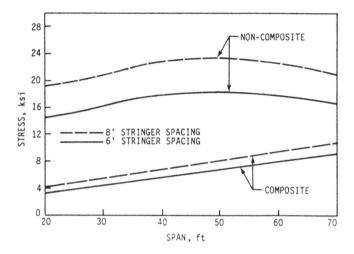

FIGURE 6.6 Stress in top flange of stringer, composite action vs. noncomposite action. (*Source:* Klaiber, F.W. et al., NCHRP 293, Transportation Research Board, 1987. With permission.)

to 40 ft (12.2 m) long, with oblong holes 12 in. (305 mm) × 4 in. (100 mm) were used to replace the existing deck. Four studs were welded to the girders through each hole. Composite action was obtained by filling the holes, as well as the gaps between the panels and steel stringers, with fast-curing concrete.

If the concrete deck does not need replacing, composite action can be obtained by coring through the existing concrete deck to the steel superstructure. Appropriate shear connectors are placed in the holes; the desired composite action is then obtained by filling the holes with nonshrink grout. This procedure was used in the reconstruction of the Pulaski Skyway near the Holland Tunnel linking New Jersey and New York [38]. After removing an asphalt overlay and some of the old concrete, the previously described procedure with welded studs placed in the holes was used. The holes were then grouted and the bridge resurfaced with latex-modified concrete.

Structural lightweight concrete has been used in both precast panels and in cast-in-place bridge decks. Comments made on normal-weight concrete in the preceding paragraphs essentially apply to lightweight concrete also. However, since the shear strength, fatigue strength, and modulus of elasticity of lightweight concrete are less than that of normal-weight concrete, these lesser values must be taken into account in design.

The advantages of composite action can be seen in Figure 6.6. Shown in this graph is the decrease in the top flange stress as a result of providing composite action on a simply supported single-span bridge with steel stringers and an 8-in. concrete deck. As may be seen in this figure, two stringer spacings, 6 ft (1.8 m) and 8 ft (2.4 m) are held constant, while the span length was varied from 20 ft (6.1 m) to 70 ft (21.3 m). These stresses are based on the maximum moment that results from either the standard truck loading (HS20-44) or the standard lane loading, whichever governs. Concrete stresses were considerably below the allowable stress limit; composite action reduced the stress in the bottom flange 15 to 30% for long and short spans, respectively. As may be seen in Figure 6.6 for a 40-ft (12.2-m) span with 8-ft (2.4-m) stringer spacing, composite action will reduce the stress in the top flange 68%, 22 ksi (152 Mpa) to 7 ksi (48 MPa). Composite action is slightly more beneficial in short spans than in long spans, and the larger the stringer spacings, the more stress reduction when composite action is added. Results for other types of deck are similar but will depend on the type and size of deck, amount of composite action obtained, type of support system, and the like.

6.3.3 Types of Shear Connectors

As previously mentioned, in order to create composite action between the steel stringers and the bridge deck some type of shear connector is required. In the past, several different types of shear connectors were used in the field; these connectors can be seen in Figure 6.7. Of these, because of the advancements

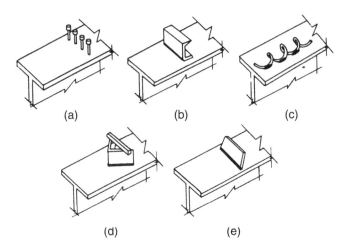

FIGURE 6.7 Common shear connectors. (a) Welded studs; (b) channel; (c) spiral; (d) stiffened angle; (e) inclined flat bar. (*Source:* Klaiber, F.W. et al., NCHRP 293, Transportation Research Board, 1987. With permission.)

and ease in application, welded studs have become the most commonly used shear connector today. In the strengthening of an existing bridge, frequently one of the older types of shear connectors will be encountered. A strength evaluation must be undertaken to ensure that the shear connectors present are adequate. The following references can be used to obtain the ultimate strength of various types of shear connectors. A method for calculating the strength of a flat bar can be found in Cook [39]; also, work done by Klaiber et al. [40] can be used in evaluating the strength of stiffened angles. Older AASHTO standard specifications can be used to obtain ultimate strength of shear connectors; for example, values for spirals can be found in the AASHTO standard specifications from 1957 to 1968. The current AASHTO specifications only give ultimate-strength equations for welded studs and channels; thus, if shear connectors other than these two are encountered, the previously mentioned references should be consulted.

The procedure employed for using high-strength bolts as shear connectors (Figure 6.8) is very similar to that used for utilizing welded studs in existing concrete, except for the required holes in the steel stringer. To minimize slip, the hole in the steel stringer is made the same size as the diameter of the bolt. Dedic and Klaiber [41] and Dallam [42, 43] have shown that the strength and stiffness of high-strength bolts are essentially the same as those of welded shear studs. Thus, existing AASHTO ultimate-strength formulas for welded stud connectors can be used to estimate the ultimate capacity of high-strength bolts.

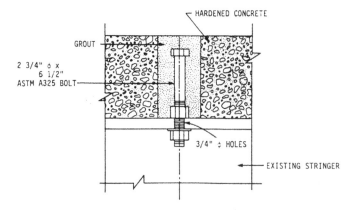

FIGURE 6.8 Details of double-nutted high-strength bolt shear connector. (*Source:* Klaiber, F.W. et al., NCHRP 293, Transportation Research Board, 1987. With permission.)

6.3.4 Design Considerations

The means of obtaining composite action will depend on the individual bridge deck. If the deck is in poor condition and needs to be replaced, the following variables should be considered: (1) weldability of steel stringers, (2) type of shear connector, and (3) precast vs. cast in place.

To determine the weldability of the shear connector, the type of steel in the stringers must be known. If the type of steel is unknown, coupons may be taken from the stringers to determine their weldability. If it is found that welding is not possible, essentially the only alternative for shear connection is high-strength bolts. Although the procedure is rarely done, bolts could be used to attach channels to the stringers for shear connection. When welding is feasible, either welded studs or channels can be used. Because of the ease of application of the welded studs, channels are rarely used today. In older constructions where steel cover plates were riveted to the beam flanges, an option that may be available is to remove the rivets connecting the top cover plate to the top flange of the beam and replace the rivets with high-strength bolts in a manner similar to that which is shown in Figure 6.8.

According to the current AASHTO manual, *Standard Specifications for Highway Bridges*, in new bridges shear connectors should be designed for fatigue and checked for ultimate strength. However, in older bridges, the remaining fatigue life of the bridge will be considerably less than that of the new shear connectors; thus, one only needs to design the new shear connectors for ultimate strength. If an existing bridge with composite action requires additional shear connectors, the ultimate strength capacity of the original shear connector (connector #1) and new shear connectors (connectors #2) can be simply added even though they are different types of connectors. Variation in the stiffness of the new shear connectors and original shear connectors will have essentially no effect on the elastic behavior of the bridge and nominal effect on the ultimate strength [44].

The most common method of creating composite action when one works with precast concrete decks is to preform slots in the individual panels. These slots are then aligned with the stringers for later placement of shear connectors (see Figure 6.5). Once shear connectors are in place, the holes are filled with nonshrink concrete. A similar procedure can be used with laminated timber except the holes for the shear connectors are drilled after the panels are placed.

When it is necessary to strengthen a continuous span, composite action can still be employed. One common approach is for the positive moment region to be designed using the same procedure as that for simple span bridges. When designing the negative moment region, the engineer has two alternatives. The engineer can continue the shear connectors over the negative moment region, in which case the longitudinal steel can be used in computing section properties in the negative moment region. The other alternative is to discontinue the shear connectors over the negative moment region. As long as the additional anchorage connectors in the region of the point of dead-load contraflexure are provided, as required by the code, continuous shear connectors are not needed. When this second alternative is used, the engineer cannot use the longitudinal steel in computing the section properties in the negative moment region. If shear connectors are continued over the negative moment region, one should check to be sure that the longitudinal steel is not overstressed. Designers should consult the pertinent AASHTO standards to meet current design guidelines.

6.4 Improving the Strength of Various Bridge Members

6.4.1 Addition of Steel Cover Plates

Steel Stringer Bridges

Description

One of the most common procedures used to strengthen existing bridges is the addition of steel cover plates to existing members. Steel cover plates, angles, or other sections may be attached to the beams by means of bolts or welds. The additional steel is normally attached to the flanges of existing sections as a means of increasing the section modulus, thereby increasing the flexural capacity of the member. In most cases the member is jacked up during the strengthening process, relieving dead-load stresses on the existing member. The new cover plate section is then able to accept both live-load and dead-load stresses when the jacks are removed, which ensures that less steel will be required in the cover plates. If the bridge is not jacked up, the cover plate will carry only live-load stresses, and more steel will be required.

Applicability, Advantages, and Disadvantages

The techniques described in this section are widely applicable to steel members whose flexural capacity is inadequate. Members in this category include steel stringers (both composite and noncomposite), floor beams, and girders on simply supported or continuous bridges. Note, however, that cover plating is most effective on composite members.

There are a number of advantages to using steel cover plates as a method of strengthening existing bridges. This method can be quickly installed and requires little special equipment and minimal labor and materials. If bottom flange stresses control the design, cover plating is effective even if the deck is not replaced. In this case, it is more effective when applied to noncomposite construction. In addition, design procedures are straightforward and thus require minimal time to complete.

In certain instances these advantages may be offset by the costly problems of traffic control and jacking of the bridge. As a minimum, the bridge may have to be closed or separate traffic lanes established to relieve any stresses on the bridge during strengthening. In addition, significant problems may develop if part of the slab must be removed in order to add cover plates to the top of the beams. When cover plates are attached to the bottom flange, the plates should be checked for underclearance if the situation requires it. Still another potential problem if welding is used is that the existing members may not be compatible with current welding materials.

The most commonly reported problem encountered with the addition of steel cover plates is fatigue cracking at the top of the welds at the ends of the cover plates. In a study by Wattar et al. [45], it was suggested that bolting be used at the cover plate ends. Tests showed that bolting the ends raises the fatigue category of the member from stress Category E to B and also results in material savings by allowing the plates to be cut off at the theoretical cutoff points.

Another method for strengthening this detail is to grind the transverse weld to a 1:3 taper [46]. This is a practice of the Maryland State Highway Department. Using an air hammer to peen the toe of the weld and introduce compressive residual stresses is also effective in strengthening the connection [46]. The fatigue strength can be improved from stress category E to D by using this technique. Either solution has been shown to reduce significantly the problem of fatigue cracking at the cover plate ends.

Materials other than flange cover plates may be added to stringer flanges for strengthening. For example, the Iowa Department of Transportation prefers to attach angles to the webs of steel I-beam bridges (either simply supported or continuous spans) with high-strength bolts as a means to reduce flexural live-load stresses in the beams. Figure 6.9 shows a project completed by the Iowa Department of Transportation involving the addition of angles to steel I-beams using high-strength bolts. In some instances the angles are attached only near the bottom flange. Normally, the bridge is not jacked up during strengthening, and only the live loads are removed from the particular I-beam being strengthened. Because the angles are bolted on, problems of fatigue cracking that could occur with welding are eliminated. This method does have one potential problem, however: the possibility of having to remove part of a web stiffener should one be crossed by an angle.

Another method of adding material to existing members for strengthening is shown in Figure 6.10 where structural Ts were bolted to the bottom flanges of the existing stringers using structural angles. This idea represented a design alternative recommended by Howard, Needles, Tammen and Bergendoff as one method of strengthening a bridge comprising three 50-ft (15.2-m) simple spans. Each of the four stringers per span was strengthened in a similar manner.

Design Procedure

The basic design steps required in the design of steel cover plates follow:

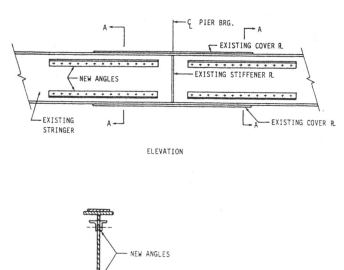

FIGURE 6.9 Iowa DOT method of adding angles to steel I-beams. (*Source:* Klaiber, F.W. et al., NCHRP 293, Transportation Research Board, 1987. With permission.)

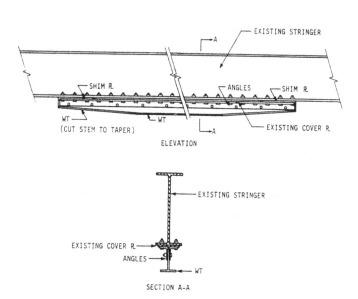

FIGURE 6.10 Strengthening of existing steel stringer by addition of structural T section. (*Source:* Klaiber, F.W. et al., NCHRP 293, Transportation Research Board, 1987. With permission.)

1. Determine moment and shear envelopes for desired live-load capacity of each beam.
2. Determine the section modulus required for each beam.
3. Determine the optimal amount of steel to achieve desired section modulus–strength requirement, fatigue requirement.
4. Design connection of cover plates to beam strength requirement, fatigue requirement.
5. Determine safe cutoff point for cover plates.

In addition to the foregoing design steps, the following construction considerations may prove helpful:

1. Grinding the transverse weld to a 1:3 taper or bolting the ends of plates rather than welding reduces fatigue cracking at the cover plate ends [46, 47].
2. In most cases a substantial savings in steel can be made if the bridge is jacked to relieve dead-load stresses prior to adding cover plates.
3. The welding of a cover plate should be completed within a working day. This minimizes the possibility of placing a continuous weld at different temperatures and inducing stress concentrations.
4. Shot blasting of existing beams to clean welding surface may be necessary.

Reinforced Concrete Bridges

Description

One method of increasing flexural capacity of a reinforced concrete beam is to attach steel cover plates or other steel shapes to the tension face of the beam. The plates or shapes are normally attached by bolting, keying, or doweling to develop continuity between the old beam and the new material. If the beam is also inadequate in shear, combinations of straps and cover plates may be added to improve both shear and flexural capacity. Because a large percentage of the load in most concrete structures is dead load, for cover plating to be most effective, the structure should be jacked prior to cover plating to reduce the dead-load stresses of the member. The addition of steel cover plates may also require the addition of concrete to the compression face of the member.

Applicability

A successful method of strengthening reinforced concrete beams has involved the attachment of a steel channel to the stem of a beam. This technique is shown in Figure 6.11. Taylor [48] performed tests on a section using steel channels and found it to be an effective method of strengthening. An advantage to this method is that rolled channels are available in a variety of sizes, require little additional preparation prior to attachment, and provide a ready formwork for the addition of grouting. The channels can also be easily reinforced with welded cover plates if additional strength is required. Prefabricated channels

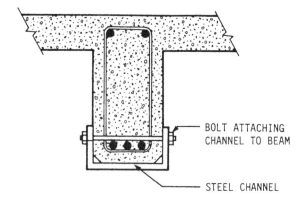

FIGURE 6.11 Addition of a steel channel to an existing reinforced concrete beam. (*Source:* Klaiber, F.W. et al., NCHRP 293, Transportation Research Board, 1987. With permission.)

are an effective substitute when rolled sections of the required size are not available. It should be noted that the bolts are placed above the longitudinal steel so that the stirrups can carry shear forces transmitted by the channels. If additional sheer capacity is required, external stirrups should also be installed. It is also recommended that an epoxy resin grout be used between the bolts and concrete. The epoxy resin grout provides greater penetration in the bolt holes, thereby reducing slippage and improving the strength of the composite action.

Bolting steel plates to the bottom and sides of beam sections has also been performed successfully, as documented by Warner [49]. Bolting may be an expensive and time-consuming method, because holes usually have to be drilled through the old concrete. Bolting is effective, however, in providing composite action between the old and new material.

The placement of longitudinal reinforcement in combination with a concrete sleeve or concrete cover is another method for increasing the flexural capacity of the member. This method is shown in Figures 6.12a and b as outlined in an article on strengthening by Westerberg [50]. Warner [49] presents a similar method that is shown in Figure 6.12c.

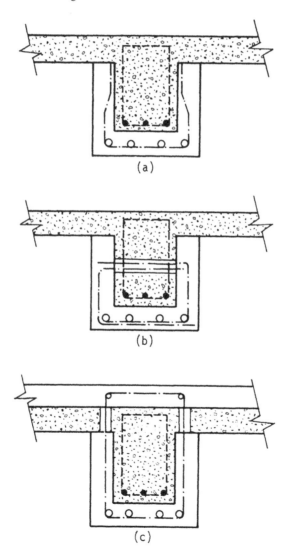

FIGURE 6.12 Techniques for increasing the flexural capacity of reinforced concrete beams with reinforced concrete sleeves. (*Source:* Klaiber, F.W. et al., NCHRP 293, Transportation Research Board, 1987. With permission.)

Developing a bond between the old and new material is critical to developing full continuity. Careful cleaning and preparation of the old concrete and the application of a suitable epoxy-resin primer prior to adding new concrete should provide adequate bonding. Stirrups should also be added to provide additional shear reinforcement and to support the added longitudinal bars.

Design and Analysis Procedure

The design of steel cover plates for concrete members is dependent on the amount of continuity assumed to exist between the old and new material. If one assumes that full continuity can be achieved and that strains vary linearly throughout the depth of the beam, calculations are basically straightforward. As stated earlier, much of the load in concrete structures is dead load, and jacking of the deck during cover plating will greatly reduce the amount of new steel required. It should also be pointed out that additional steel could lead to an overreinforced section. This could be compensated for by additional concrete or reinforcing steel in the compression zone.

Case Studies

Steel cover plates can be used in a variety of situations. They can be used to increase the section modulus of steel, reinforced concrete, and timber beams. Steel cover plates are also an effective method of strengthening compression members in trusses by providing additional cross-sectional area and by reducing the slenderness ratio of the member.

Mancarti [51] reported the use of steel cover plates to strengthen floor beams on the Pit River Bridge and Overhead in California. The truss structure required strengthening of various other components to accommodate increased dead load. Stringers in this bridge were strengthened by applying prestressing tendons near the top flange to reduce tensile stress in the negative moment region. This prestressing caused increased compressive stresses in the bottom flanges, which in turn required the addition of steel bars to the tops of the stringer bottom flange.

In a report by Rodriguez et al. [52], a number of cases of cover-plating existing members of old railway trusses were cited. These case studies included the inspection of 109 bridges and a determination of their safety. Some strengthening techniques included steel-cover-plating beam members as well as truss members. Cover plates used to reinforce existing floor beams on a deficient through-truss were designed to carry all live-load bending moment. Deficient truss members were strengthened with box sections made up of welded plates. The box was placed around the existing member and connected to it by welding.

6.4.2   Shear Reinforcement

External Shear Reinforcement for Concrete, Steel, and Timber Beams

The shear strength of reinforced concrete beams or prestressed concrete beams can be improved with the addition of external steel straps, plates, or stirrups. Steel straps are normally wrapped around the member and can be post-tensioned. Post-tensioning allows the new material to share both dead and live loads equally with the old material, resulting in more efficient use of the material added. A disadvantage of adding steel straps is that cutting the deck to apply the straps leaves them exposed on the deck surface and thus difficult to protect. By contrast, adding steel plates does not require cutting through the deck. The steel plates are normally attached to the beam with bolts or dowels.

External stirrups may also be applied with different configurations. Figure 6.13a shows a method of attaching vertical stirrups using channels at the top and bottom of the beam. The deck (not shown in either figure) provides protection for the upper steel channel [53]. Adding steel sections at the top of the beam web and attaching stirrups is shown in Figure 6.13b. In this manner, cutting holes through the deck is eliminated. External stirrups can also be post-tensioned in most situations if desired.

Another method of increasing shear strength is shown in Figure 6.14. This method is a combination of post-tensioning and the addition of steel in the form of prestressing tendons. As recommended in a strengthening manual by the OECD [54], tendons may be added in a vertical or inclined orientation and may be placed either within the beam web or inside the box as shown in the figure. Care should be taken to avoid overstressing parts of the structure when prestressing. If any cracks exist in the member, it is a

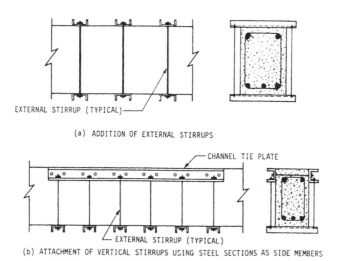

FIGURE 6.13 Methods of adding external shear reinforcement to reinforced concrete beams. (*Source:* Klaiber, F.W. et al., NCHRP 293, Transportation Research Board, 1987. With permission.)

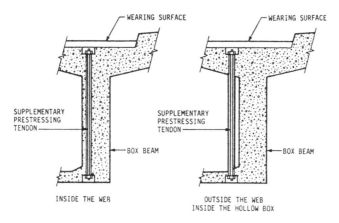

FIGURE 6.14 External shear reinforcement of box beam girders. (*Source:* Klaiber, F.W. et al., NCHRP 293, Transportation Research Board, 1987. With permission.)

good practice to inject them with an epoxy before applying the prestressing forces. Documentation of this type of reinforcement technique is made also by Suter and Audrey [55] and Dilger and Ghali [56]. Figure 6.15 illustrates the technique used by Dilger and Ghali [56] where web thickening was added to the inside of the box web before adding external reinforcement consisting of stressed steel bars. The thickening was required to reduce calculated tensile stresses at the outside of the web due to prestressing the reinforcement.

West [57] makes reference to a number of methods of attaching steel plates to deficient steel I-beam girder webs as a means of increasing their shear strength. The steel plates are normally of panel size and are attached between stiffeners by bolting or welding. Where shear stresses are high, the plates should fit tightly between the stiffeners and girder flanges. West indicates that one advantage of this method is that it can be applied under traffic conditions.

Timber stringers with inadequate shear capacity can be strengthened by adding steel cover plates. NCHRP Report 222 [11] demonstrates a method of repairing damaged timber stringers with inadequate shear capacity. The procedure involves attaching steel plates to the bottom of the beam in the deficient region and attaching it with draw-up bolts placed on both sides of the beam. Holes are drilled through the top of the deck, and a steel strap is placed at the deck surface and at the connection to the bolts.

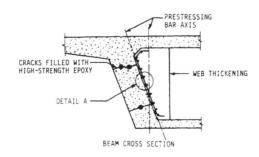

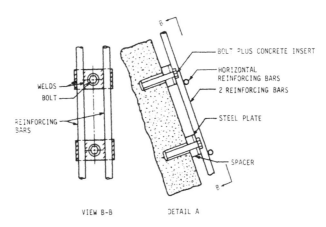

FIGURE 6.15 Details of web reinforcement to strengthen box beam in shear. (*Source:* Klaiber, F.W. et al., NCHRP 293, Transportation Research Board, 1987. With permission.)

Epoxy Injection and Rebar Insertion

The Kansas Department of Transportation has developed and successfully used a method for repairing reinforced concrete girder bridges. The bridges had developed shear cracks in the main longitudinal girders [58]. The procedure used by the Kansas Department of Transportation not only prevented further shear cracking but also significantly increased the shear strength of the repaired girders.

The method involves locating and sealing all of the girder cracks with silicone rubber, marking the girder centerline on the deck, locating the transverse deck reinforcement, vacuum drilling 45° holes that avoid the deck reinforcement, pumping the holes and cracks full of epoxy, and inserting reinforcing bars into the epoxy-filled holes. A typical detail is shown in Figure 6.16.

An advantage of using the epoxy repair and rebar insertion method is its wide application to a variety of bridges. Although the Kansas Department of Transportation reported using this strengthening method on two-girder, continuous, reinforced concrete bridges, this method can be a practical solution on most types of prestressed concrete beam and reinforced concrete girder bridges that require additional shear strength. The essential equipment requirements needed for this strengthening method may limit its usefulness, however. Prior to drilling, the transverse deck steel must be located. The drilling unit and vacuum pump required must be able to quickly drill straight holes to a controlled depth and keep the holes clean and free of dust.

Addition of External Shear Reinforcement

Strengthening a concrete bridge member that has a deficient shear capacity can be performed by adding external shear reinforcement. The shear reinforcement may consist of steel side plates or steel stirrup reinforcement. This method has been applied on numerous concrete bridge systems.

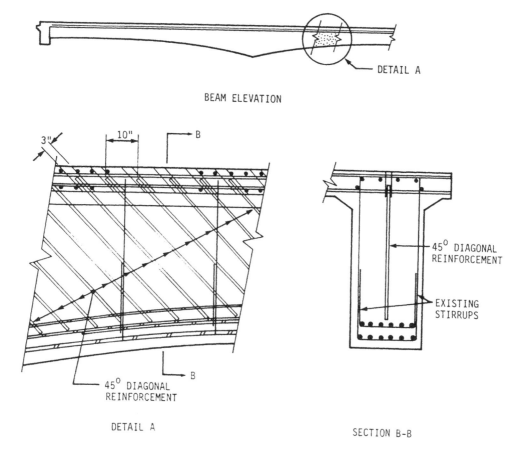

FIGURE 6.16 Kansas DOT shear strengthening procedure. (*Source:* Klaiber, F.W. et al., NCHRP 293, Transportation Research Board, 1987. With permission.)

A method proposed by Warner [49] involves adding external stirrups. The stirrups consist of steel rods placed on both sides of the beam section and attached to plates at the top and bottom of the section. In some applications, channels are mounted on both sides at the top of the section to attach the stirrups. This eliminates drilling through the deck to make the connection to a plate.

In a study by Dilger and Ghali [56], external shear reinforcement was used to repair webs of prestressed concrete bridges. Although the measures used were intended to bring the deficient members to their original flexural capacity, the techniques applied could be used for increasing the shear strength of existing members. Continuous box girders in the 827-ft (252-m) long bridges had become severely cracked when prestressed. The interior box beam webs were strengthened by the addition of 1-in. (25-mm) diameter steel rods placed on both sides of the web. Holes were drilled in the upper and lower slabs as close as possible to the web to minimize local bending stresses in the slabs. Post-tensioning tendons were placed through the holes, stressed, and then anchored.

The slanted outside webs were strengthened with reinforcing steel. Before the bars were added, the inside of the web was "thickened" and the reinforcement was attached with anchor bolts placed through steel plates that were welded to the reinforcement. The web thickening was necessary because the prestressing would have produced substantial tensile stresses at the outside face of the web.

6.4.3 Jacketing of Timber or Concrete Piles and Pier Columns

Improving the strength of timber or concrete piles and pier columns can be achieved by encasing the column in concrete or steel jackets. The jacketing may be applied to the full length of the column or

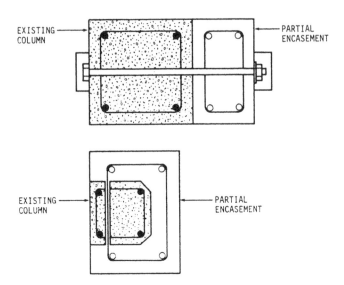

FIGURE 6.17 Partial jacketing of an existing column. (*Source:* Klaiber, F.W. et al., NCHRP 293, Transportation Research Board, 1987. With permission.)

only to severely deteriorated sections. The jacketing increases the cross-sectional area of the column and reduces the slenderness ratio of the column. Partial encasement of a column can also be particularly effective when an unbalanced moment acts on the column. Figure 6.17 illustrates two such concepts for member addition that were noted from work on strengthening reinforced concrete structures in Europe [50].

Completely encasing the existing column in a concrete jacket has been a frequently used method of strengthening concrete pier columns. Normally, the reinforcement is placed around the existing column perimeter inside the jacket and "ramset" to the existing member [50]. The difficulty most often observed with this technique is developing continuity between the old and new material. This is critical if part of the load is to be transferred to the new material. Work by Soliman [59] on repair of reinforced concrete columns by jacketing has included an experimental investigation of the bond stresses between the column and jacket. The first step is normally surface preparation of the existing concrete column. Consideration should also be given at this time to jacking of the superstructure and placing temporary supports on either side of the column. Soliman [59] concludes that this is an important step, since the shrinkage phenomenon causes compressive stresses on the column that will be reduced if the existing column is unloaded. In addition, supports will be necessary if the column shows significant signs of deterioration. This procedure will also allow the new material to share equally both dead and live loads after the supports are removed. Additional longitudinal reinforcing bars and stirrups are then placed around the column. Spiral stirrup reinforcement should be used because it will provide greater strength and ductility than normal stirrups [59]. An epoxy resin is then applied to the old concrete to increase the bonding action between the old concrete and the concrete to be added. Formwork is then erected to form the jacket, and concrete is placed and compacted.

Jacketing techniques have been used extensively for seismic retrofitting of existing pier columns. A recent report by Wipf et al. [60] provides an extensive list and discussion of various retrofit methods for reinforced concrete bridge columns, including the use of steel jackets and fiber-reinforced polymer wraps.

Modification Jacketing

Increasing the load-carrying capacity of bridge pier columns or timber piles supporting bent caps is normally achieved through the addition of material to the existing cross section. Jacketing or adding a sleeve around the column perimeter can be performed a number of ways.

In a paper by Karamchandani [61], various concepts for jacketing existing members are illustrated. These include addition of reinforcement and concrete around three sides of rectangular beams as well as placement only at the bottom of the beam web. Additional schemes are also illustrated for column members. The effectiveness of this method depends on the degree of adhesion between new and existing concrete, which can vary between 30 and 80% of the total strength of the *in situ* concrete. The author suggests welding new reinforcing to the existing reinforcement and using concrete with a slump of 3 to 4 in. (75 to 100 mm). The use of rapid hardening cements is not recommended, since it results in a lower strength of concrete on the contact surface because of high contraction stresses.

The addition of concrete collars on reinforced concrete columns is performed most efficiently by using circular reinforcement rather than dowels or shear keys according to Klein and Gouwens [62]. While the other methods may require costly and time-consuming drilling and/or cutting, circular reinforcement does not. When this method is used, shear-friction is the primary load-transfer mechanism between the collar and the existing column. Klein and Gouwens have outlined a design procedure for this strengthening method.

In a paper by Syrmakezis and Voyatzis [63], an analytical method for calculating the stiffness coefficients of columns strengthened by jacketing is presented. The procedure uses compatibility conditions for the deformations of the strengthened system and the analysis can consider rigid connections between the jacket and column on a condition where relative slip is allowed.

6.5 Post-Tensioning Various Bridge Components

6.5.1 Introduction

Since the 19th century, timber structures have been strengthened by means of king post and queen post-tendon arrangements; these forms of strengthening by post-tensioning are still used today. Since the 1950s, post-tensioning has been applied as a strengthening method in many more configurations to almost all common bridge types. The impetus for the recent surge in post-tensioning strengthening is undoubtedly a result of its successful history of more than 40 years and the current need for strengthening of bridges in many countries.

Post-tensioning can be applied to an existing bridge to meet a variety of objectives. It can be used to relieve tension overstresses with respect to service load and fatigue-allowable stresses. These overstresses may be axial tension in truss members or tension associated with flexure, shear, or torsion in bridge stringers, beams, or girders.

Post-tensioning also can reduce or reverse undesirable displacements. These displacements may be local, as in the case of cracking, or global, as in the case of excessive bridge deflections. Although post-tensioning is generally not as effective with respect to ultimate strength as with respect to service-load-allowable stresses, it can be used to add ultimate strength to an existing bridge. It is possible to use post-tensioning to change the basic behavior of a bridge from a series of simple spans to continuous spans. All of these objectives have been fulfilled by post-tensioning existing bridges, as documented in the engineering literature.

Most often, post-tensioning has been applied with the objective of controlling longitudinal tension stresses in bridge members under service-loading conditions. Figure 6.18 illustrates the axial forces, shear forces, and bending moments that can be achieved with several simple tendon configurations. The concentric tendon in Figure 6.18a will induce an axial compression force that, depending on magnitude, can eliminate part or all of an existing tension force in a member or even place a residual compression force sufficient to counteract a tension force under other loading conditions. The amount of post-tensioning force that can safely be applied, of course, is limited by the residual-tension dead-load force in the member.

The tendon configuration in Figure 6.18a is generally used only for tension members in trusses, whereas the remaining tendon configurations in Figure 6.18 would be used for stringers, beams, and girders. The eccentric tendon in Figure 6.18b induces both axial compression and negative bending. The eccentricity

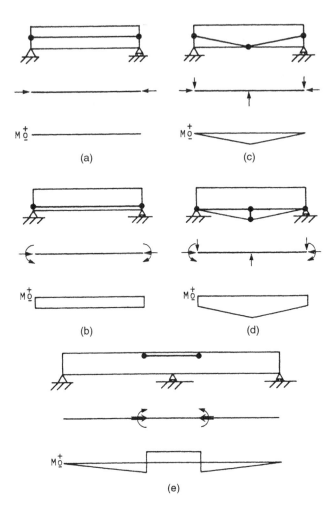

FIGURE 6.18 Forces and moment induced by longitudinal post-tensioning. (a) Concentric tendon; (b) eccentric tendon; (c) polygonal tendon; (d) king post; (e) eccentric tendon, two-span member. (*Source:* Klaiber, F.W. et al., NCHRP 293, Transportation Research Board, 1987. With permission.)

of the tendon may be varied to control the proportions of axial compression vs. bending applied to the member. Length of the tendon also may be varied to apply post-tensioning only to the most highly stressed portion of the member. The polygonal tendon profile in Figure 6.18c also induces axial compression and negative bending, but the negative bending is nonuniform within the post-tensioned region. Locations of bends on the tendon and eccentricities of the attachments at the bends can be set to control the moments caused by the post-tensioning. The polygonal tendon also induces shear forces that are opposite to those applied by live and dead loads.

The king post tendon configuration in Figure 6.18d is a combination of the eccentric and polygonal tendon configurations. Because the post is beyond the profile of the original member, the proportion of moment to axial force induced in the member to be strengthened will be large.

The tendon configuration in Figure 6.18e is an eccentric tendon attached over the central support of a two-span member. In this configuration, the amount of positive moment applied in the central support region depends not only on the force in the tendon and its eccentricity, but also on the locations of the anchorages on the two spans. If the anchorages are moved toward the central support, the amount of positive moment applied will be greater than if the anchorages are moved away from the central support. This fact and the fact that there is some distribution of moment and force among parallel post-tensioned members have not always been correctly recognized, and there are published errors in the literature.

The axial force, shear force, and bending moment effects of post-tensioning described above have enough versatility in application so as to meet a wide variety of strengthening requirements. Probably this is the only strengthening method that can actually reverse undesirable behavior in an existing bridge rather than provide a simple patching effect. For both these reasons, post-tensioning has become a very commonly used repair and strengthening method.

6.5.2 Applicability and Advantages

Post-tensioning has many capabilities: to relieve tension, shear, bending, and torsion-overstress conditions; to reverse undesirable displacements; to add ultimate strength; to change simple span to continuous span behavior. In addition, post-tensioning has some very practical advantages. Traffic interruption is minimal; in some cases, post-tensioning can be applied to a bridge with no traffic interruption. Few site preparations, such as scaffolding, are required. Tendons and anchorages can be prefabricated. Post-tensioning is an efficient use of high strength steel. If tendons are removed at some future date, the bridge will generally be in no worse condition than before strengthening.

To date, post-tensioning has been used to repair or strengthen most common bridge types. Most often, post-tensioning has been applied to steel stringers, floor beams, girders, and trusses, and case histories for strengthening of steel bridges date back to the 1950s. Since the 1960s, external post-tensioning has been applied to reinforced concrete stringer and T bridges. In the past 20 years, external post-tensioning has been added to a variety of prestressed, concrete stringer and box beam bridges. Many West German prestressed concrete bridges have required strengthening by post-tensioning due to construction joint distress. Post-tensioning even has been applied to a reinforced concrete slab bridge by coring the full length of the span for placement of tendons [63].

Known applications of post-tensioning will be idealized and summarized as Schemes A through L in Figures 6.19 through 6.22. Typical schemes for stringers, beams, and girders are contained in Figure 6.19. The simplest and, with the exception of the king post, the oldest scheme is Scheme A: a straight, eccentric tendon shown in Figure 6.19a. Lee [64] reported use of the eccentric tendon for strengthening of British cast iron and steel highway and railway bridges in the early 1950s. Since then, Scheme A has been applied to many bridges in Europe, North America, and other parts of the world. Scheme A is most efficient if the tendon has a length less than that of the member, so that the full post-tensioning negative moment is not applied to regions with small dead-load moments. The variation on Scheme A for continuous spans, Scheme AA in Figure 6.19e, has been reported in use for deflection control or strengthening in Germany [65] and the United States [66] since the late 1970s.

The polygonal tendon, Scheme B in Figure 6.19b and its extension to continuous spans, Scheme BB in Figure 6.19f, has been in use since at least the late 1960s. Vernigora et al. [67] reported the use of Scheme BB for a five-span, reinforced-concrete T-beam bridge in 1969. The bridge over the Welland Canal in Ontario, Canada, was converted from simple-span to continuous-span behavior by means of external post-tensioning cables.

Scheme C in Figure 6.19c provided the necessary strengthening for a steel plate, girder railway bridge in Czechoslovakia in 1964 [68]. The tendons and compression struts for the bridge were fabricated from steel T sections, and the tendons were stressed by deflection at bends rather than by elongation as is the usual case. The tendons for the plate girder bridge were given a three-segment profile to apply upward forces at approximately the third points of the span, so that the existing dead-load moments could be counteracted efficiently. In the late 1970s in the United States, Kandall [69] recommended use of Scheme C for strengthening because it does not place additional axial compression in the existing structure. For other schemes, the additional axial compression induced by post-tensioning will add compressive stress to regions that may be already overstressed in compression.

Scheme D in Figure 6.19d was used in Minnesota in 1975 to strengthen temporarily a steel stringer bridge [70]. It was possible to strengthen that bridge economically with scrap timber and cable for the last few years of its life before it was replaced.

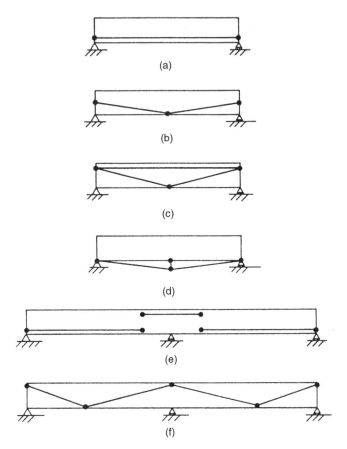

FIGURE 6.19 Tendon configurations for flexural post-tensioning of beams. (a) Scheme A, eccentric tendon; (b) Scheme B, polygonal tenton; (c) Scheme C, polygonal tendon with compression strut; (d) Scheme D, king post; (e) Scheme AA, eccentric tendons; (f) Scheme BB, polygonal tendons. (*Source:* Klaiber, F.W. et al., NCHRP 293, Transportation Research Board, 1987. With permission.)

The tendon schemes in Figure 6.19, in general, appear to be very similar to reinforcing bar patterns for concrete beams. Thus, it is not surprising that post-tensioning also has been used for shear strengthening, in patterns very much like those for stirrups in reinforced concrete beams. Scheme E in Figure 6.20a illustrates a pattern of external stirrups for a beam in need of shear strengthening. Types of post-tensioned external stirrups have been used or proposed for timber beams [11], reinforced concrete beams and, as illustrated in Figure 6.20b, for prestressed concrete box-girder bridges [71].

Post-tensioning was first applied to steel trusses for purposes of strengthening in the early 1950s [64], at about the same time that it was first applied to steel stringer and steel girder, floor beam bridges. Typical strengthening schemes for trusses are presented in Figure 6.21. Scheme F, concentric tendons on individual members, shown in Figure 6.21a, was first reported for the proposed strengthening of a cambered truss bridge in Czechoslovakia in 1964 [68]. For that bridge it was proposed to strengthen the most highly stressed tension diagonals by post-tensioning. Scheme F tends to be uneconomical because it requires a large number of anchorages, and very few truss members benefit from the post-tensioning.

Scheme G in Figure 6.21b, a concentric tendon on a series of members, has been the most widely used form of post-tensioning for trusses. Lee [64] describes the use of this scheme for British railway bridges in the early 1950s, and there have been a considerable number of bridges strengthened with this scheme in Europe.

The polygonal tendon in Scheme H, Figure 6.21c, has not been reported for strengthening purposes, but it has been used in the continuous-span version of Scheme I in Figure 6.21d for a two-span truss

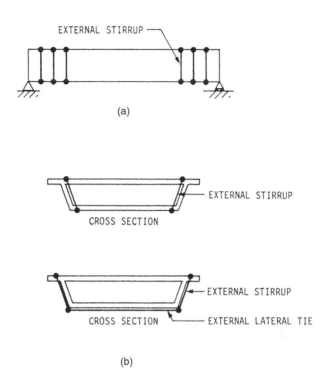

FIGURE 6.20 Tendon configurations for shear post-tensioning. (*Source:* Klaiber, F.W. et al., NCHRP 293, Transportation Research Board, 1987. With permission.)

bridge in Switzerland [72]. In the late 1960s, a truss highway bridge in Aarwangen, Switzerland, was strengthened by means of four-segment tendons on each of the two spans. The upper chord of each truss was unable to carry the additional compression force induced by the post-tensioning, and, therefore, a free-sliding compression strut was added to each top chord to take the axial post-tensioning force.

Scheme J, the king post in Figure 6.21e, has been suggested for new as well as existing trusses [7]; however, cases of its actual use for strengthening have not been reported in the literature. Because most trusses are placed on spans greater than 100 ft (30.5 m), the posts below the bridge could extend down quite far and severely reduce clearance under the bridge. The king post or queen post would thus be in a very vulnerable position and would not be appropriate in many situations.

Most uses of post-tensioning for strengthening have been on the longitudinal members in bridges; however, post-tensioning has also been used for strengthening in the transverse direction. After the deterioration of the lateral load distribution characteristics of laminated timber decks was noted in Canada in the mid-1970s [73], Scheme K in Figure 6.22a was used to strengthen the deck. A continuous-steel channel waler at each edge of the deck spreads the post-tensioning forces from threadbar tendons above and below the deck, thereby preventing local overstress in the timber. A similar tendon arrangement, Scheme L in Figure 6.22b, was used in an Illinois bridge [74] to tie together spreading, prestressed concrete box beams.

The overview of uses of post-tensioning for bridge strengthening given above identifies the most important concepts that have been used in the past and indicates the versatility of post-tensioning as a strengthening method.

6.5.3 Limitations and Disadvantages

When post-tensioning is used as a strengthening method, it increases the allowable stress range by the magnitude of the applied post-tensioning stress. If maximum advantage is taken of the increased allowable-stress range, the factor of safety against ultimate load will be reduced. The ultimate-load capacity thus will not increase at the same rate as the allowable-stress capacity. For short-term strengthening

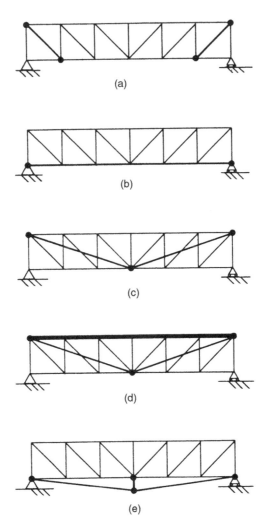

FIGURE 6.21 Tendon configurations for post-tensioning trusses. (a) Scheme F, concentric tendons on individual members; (b) Scheme G, concentric tendon on a series of members; (c) Scheme H, polygonal tendon; (d) Scheme I, polygonal tendon with compression strut; (e) Scheme J, king post. (*Source:* Klaiber, F.W. et al., NCHRP 293, Transportation Research Board, 1987. With permission.)

applications, the reduced factor of safety should not be a limitation, especially in view of the recent trend toward smaller factors of safety in design standards. For long-term strengthening applications, however, the reduced factor of safety may be a limitation.

At anchorages and brackets where tendons are attached to the bridge structure, there are high local stresses that require consideration. Any cracks initiated by holes or expansion anchors in the structure will spread with live-load dynamic cycling.

Because post-tensioning of an existing bridge affects the entire bridge (beyond the members that are post-tensioned), consideration must be given to the distribution of the induced forces and moments within the structure. If all parallel members are not post-tensioned, if all parallel members are not post-tensioned equally, or if all parallel members do not have the same stiffness, induced forces and moments will be distributed in some manner different from what is assumed in a simple analysis.

Post-tensioning does require relatively accurate fabrication and construction and relatively careful monitoring of forces locked into the tendons. Either too much or too little tendon force can cause overstress in the members of the bridge being strengthened.

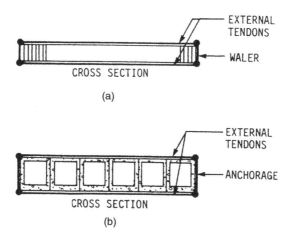

FIGURE 6.22 Tendon configuration for transverse post-tensioning of decks. (a) Scheme K, concentric tendons and walers, laminated timber deck; (b) Scheme L, concentric tendons, box beams. (*Source:* Klaiber, F.W. et al., NCHRP 293, Transportation Research Board, 1987. With permission.)

Tendons, anchorages, and brackets require corrosion protection because they are generally in locations that can be subjected to saltwater runoff or salt spray. If tendons are placed beyond the bridge profile, they are vulnerable to damage from overheight vehicles passing under the bridge or vulnerable to damage from traffic accidents. Exposed tendons also are vulnerable to damage from fires associated with traffic accidents.

6.5.4 Design Procedures

In general, strengthening of bridges by post-tensioning can follow established structural analysis and design principles. The engineer must be cautious, however, in applying empirical design procedures as they are established only for the conditions of a particular strengthening problem.

Every strengthening problem requires careful examination of the existing structure. Materials in an existing bridge were produced to some previous set of standards and may have deteriorated due to exposure over many years. The existing steel in steel members may not be weldable with ordinary procedures, and steel shapes are not likely to be dimensioned to current standards. Shear connectors and other parts may have unknown capacities due to unusual configurations.

Strengthening an existing bridge involves more than strengthening individual members. Even a simple-span bridge is indeterminate, and post-tensioning and other strengthening will affect the behavior of the entire bridge. If the indeterminate nature of the bridge is not recognized during analysis, the post-tensioning applied for strengthening purposes may not have the desired stress-relieving effects and may actually cause overstress.

Post-tensioning involves application of relatively large forces to regions of a structure that were not designed for such large forces. There is more likelihood of local overstress at tendon anchorages and brackets than at conventional member connections. Brackets need to be designed to distribute the concentrated post-tensioning forces over sufficiently large portions of the existing structure.

Members and bridges subjected to longitudinal post-tensioning will shorten axially and, depending on the tendon configuration, also will shorten and elongate with flexural stresses. These shortening and elongation effects must be considered, so that the post-tensioning has its desired effect. Frozen bridge bearings require repair and lubrication, and support details should be checked for restraints.

External tendons, whether cable or threadbar, are relatively vulnerable to corrosion, damage from overheight vehicles, traffic accidents, or fires associated with accidents. Corrosion protection and placement of the tendons are thus very important with respect to the life of the post-tensioning. Safety is also a consideration because a tendon that ruptures suddenly can pose a hazard.

For the past few years, the authors and other Iowa State University colleagues have been investigating the use of external post-tensioning (Scheme A and AA in Figure 6.19) for strengthening existing single-span and continuous-span steel stringer bridges. The research, which has been recently completed, involved laboratory testing, field implementation, and the development of design procedures. The strengthening procedures that were developed are briefly described in the following sections.

6.5.5   Longitudinal Post-Tensioning of Stringers

Simple Spans

Essentially all single-span composite steel stringer bridges constructed in Iowa between 1940 and 1960 have smaller exterior stringers. These stringers are significantly overstressed for today's legal loads; interior stringers are also overstressed to a lesser degree. Thus, the post-tension system developed is only applied to the exterior stringers; through lateral load distribution a stress reduction is also obtained in the interior stringers.

By analyzing an undercapacity bridge, an engineer can determine the overstress in the interior and exterior stringers. This overstress is based on the procedure of isolating each bridge stringer from the total structure. The amount of post-tensioning required to reduce the stress in the stringers can then be determined if the amount of post-tensioning force remaining on the exterior stringers is known; this force can be quantified with force and moment fractions. A force fraction, FF, is the ratio of the axial force that remains on a post-tensioned stringer at midspan to the sum of the axial forces for all bridge stringers at midspan, while a moment fraction, MF, is the moment remaining on the post-tensioned stringer divided by the sum of midspan moments for all bridge stringers. Knowing these fractions, the required post-tensioning force may be determined by utilizing the following relationship:

$$f = FF\left[\frac{P}{A}\right] + MF\left[\frac{Pec}{I}\right] \tag{6.1}$$

where

f = desired stress reduction in stringer lower flange
P = post-tensioning force required on each exterior stringer
A = cross-sectional area of exterior stringers
e = eccentricity of post-tensioning force measured from the neutral axis of the bridge
c = distance from neutral axis of stringer to lower flange
I = moment of inertia of exterior stringer at section being analyzed

Force fractions and moment fractions as well as other details on the procedure may be found in reference [75].

Span length and relative beam stiffness were determined to be the most significant variables in the moment fractions. As span length increases, exterior beams retain less moment; exterior beams that are smaller than the interior beams retain less post-tensioning moment than if the beams were all the same size.

The strengthening procedure and design methodology just described have been used on several bridges in the states of Iowa, Florida, and South Dakota. In all instances, the procedure was employed by local contractors without any significant difficulties. Application of this strengthening procedure to a 72-ft (34.0-m) long 45° skewed bridge in Iowa is shown in Figure 6.23.

Continuous Spans

Similar to the single-span bridges, Iowa has a large number of continuous-span composite steel stringer bridges that also have excessive flexural stresses. Through laboratory tests, it was determined that the desired stress reduction could be obtained by post-tensioning the positive moment regions of the various stringers in most situations. In the cases in which there are excessive overstresses in the negative moment regions, it may be necessary to use superimposed trusses (see Figure 6.24) on the exterior stringers in addition to post-tensioning the positive moment regions. Similar to single-span bridges, it was decided to use force fractions and moment fractions to determine the distribution of strengthening forces in a

FIGURE 6.23 Single-span bridge strengthened by post-tensioning.

given bridge. As one would expect, the design procedure is considerably more involved for continuous-span bridges as one has to consider transverse and longitudinal distribution of forces.

The required strengthening forces and final stringer envelopes should be calculated. The various strengthening schemes that can be used are shown in Figure 6.25. A designer selects the schemes required for obtaining the desired stress reduction. For additional details on the strengthening procedure the reader is referred to reference [76]. Shown in Figure 6.26 is a three-span continuous bridge near Mason City, Iowa, that has been strengthened using the schemes shown in Figure 6.25.

6.6 Developing Additional Bridge Continuity

6.6.1 Addition of Supplemental Supports

Description

Supplemental supports can be added to reduce span length and thereby reduce the maximum positive moment in a given bridge. By changing a single-span bridge to a continuous, multiple-span bridge, stresses in the bridge can be altered dramatically, thereby improving the maximum live-load capacity of the bridge. Even though this method may be quite expensive because of the cost of adding an additional pier(s), it may still be desirable in certain situations.

Applicability and Advantages

This method is applicable to most types of stringer bridges, such as steel, concrete, and timber, and has also been used on truss bridges [7]. Each of these types of bridges has distinct differences.

If a supplemental center support is added to the center of an 80-ft (24.4-m) long steel stringer bridge that has been designed for HS20-44 loading, the maximum positive live-load moment is reduced from 1164.9 ft-kips (1579.4 kN·m) to 358.2 ft-kips (485.7 kN·m), which is a reduction of over 69%. At the

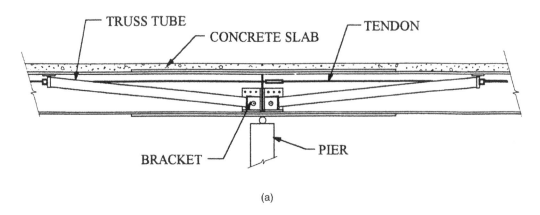

(a)

(b)

FIGURE 6.24 Superimposed truss system. (a) Superimposed truss; (b) photograph of superimposed truss.

same time, however, a negative moment of 266.6 ft-kips (361.5 kN·m) is created which must be taken into account. In situations where the added support cannot be placed at the center, reductions in positive moments are slightly less.

Limitations and Disadvantages

Depending on the type of bridge, there are various limitations in this method of strengthening. First, because of conditions directly below the existing bridge, there may not be a suitable location for the pier, as, for example, when the bridge requiring strengthening passes over a roadway or railroad tracks. Other constraints, such as soil conditions, the presence of a deep gorge, or stream velocity, could greatly increase the length of the required piles, making the cost prohibitive.

This method is most cost-effective with medium- to long-span bridges. This eliminates most timber stringer bridges because of their short lengths. In truss bridges, the trusses must be analyzed to determine the effect of adding an additional support. All members would have to be examined to determine if they could carry the change in force caused by the new support. Of particular concern would be members originally designed to carry tension, but which because of the added support must now carry compressive stresses. Because of these problems, the emphasis in this section will be on steel and concrete stringer bridges.

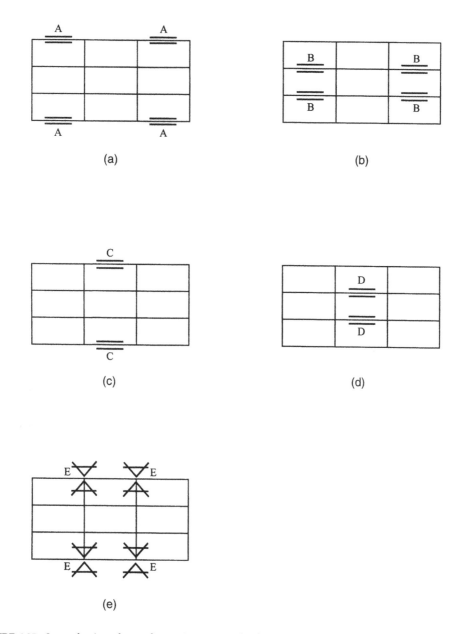

FIGURE 6.25 Strengthening schemes for continuous-span bridge. (a) Strengthening Scheme A: post-tensioning end spans of the exterior stringers; (b) strengthening Scheme B: post-tensioning end spans of the interior stringers; (c) strengthening Scheme C: post-tensioning center spans of the exterior stringers; (d) strengthening Scheme D: post-tensioning center spans of the interior strangers; (e) strengthening Scheme E: superimposed trusses at the piers of the exterior stringers.

Design Considerations

Because the design of each intermediate pier system is highly dependent on many variables such as the load on pier, width and height of bridge, and soil conditions, it is not feasible to include a generalized design procedure for piers. The engineer should use standard pier design procedures. A brief discussion of several of the more important considerations (condition of the bridge, location of pier along bridge, soil condition, type of pier, and negative moment reinforcement) is given in the following paragraphs.

FIGURE 6.26 Photograph of three-span continuous bridge strengthened with post-tensioning and superimposed trusses.

Providing supplemental support is quite expensive; therefore, the condition of the bridge is very important. If the bridge is in good to excellent condition and the only major problem is that the bridge lacks sufficient capacity for present-day loading, this method of strengthening should be considered. On the other hand, if the bridge has other deficiencies, such as a badly deteriorated deck or insufficient roadway width, a less expensive strengthening method with a shorter life should be considered.

The type of pier system employed greatly depends on the loading and also the soil conditions. The most common type of pier system used in this method is either steel H piles or timber piles with a steel or timber beam used as a pier cap. A method employed by the Florida Department of Transportation [77] can be used to install the piles under the bridge with limited modification to the existing bridge. This method consists of cutting holes through the deck above the point of application of the piles. Piles are then driven into position through the deck. The piles are then cut off so that a pier cap and rollers can be placed under the stringers. Other types of piers, such as concrete pile bents, solid piers, or hammerhead piers, can also be used; however, cost may restrict their use.

Another major concern with this method is how to provide reinforcement in the deck when the region in the vicinity of the support becomes a negative moment region. With steel stringers the bridge may either be composite or noncomposite. If noncomposite, the concrete deck is not required to carry any of the negative moment and therefore needs no alteration. On the other hand, if composite action exists, the deck in the negative moment region should be removed and replaced with a properly reinforced deck. For concrete stringer bridges the deck in the negative moment region should be removed. Reinforcement to ensure shear connection between the stringers and deck must be installed and the deck replaced with a properly reinforced deck. This method, although expensive and highly dependent on the surroundings, may be quite effective in the right situation.

6.6.2 Modification of Simple Spans

Description

In this method of strengthening, simply supported adjacent spans are connected together with a moment and shear-type connection. Once this connection is in place, the simple spans become one continuous span, which alters the stress distribution. The desired decrease in the maximum positive moment, however, is accompanied by the development of a negative moment over the interior supports.

Applicability and Advantages

This method can be used primarily with steel and timber bridges. Although it could also be used on concrete stringer bridges, the difficulties in structural connecting to adjacent reinforced concrete beams result in the method being impractical. The stringer material and the type of deck used will obviously dictate construction details. Thus, the main advantage of this procedure is that it is possible to reduce positive moments (obviously the only moments present in simple spans) by working over the piers and not near the midspan of the stringers. This method also reduces future maintenance requirements because it eliminates a roadway joint and one set of bearings at each pier where continuity is provided [12].

Limitations and Disadvantages

The main disadvantage of modifying simple spans is the negative moment developed over the piers. To provide continuity, regardless of the type of stringers or deck material, one must design for and provide reinforcement for the new negative moments and shears. Providing continuity also increases the vertical reactions at the interior piers; thus, one must check the adequacy of the piers to support the increase in axial load.

Design Considerations

The main design consideration for both types of stringers (steel and timber) concerns how to ensure full connection (shear and moment) over the piers. The following sections will give some insight into how this may be accomplished.

Steel Stringers

Berger [12] has provided information, some of which is summarized here, on how to provide continuity in a steel stringer concrete deck system. If the concrete deck is in sound condition, a portion of it must be removed over the piers. A splice, which is capable of resisting moment as well as shear, is then installed between adjacent stringers. Existing bearings are removed and a new bearing assembly is installed. In most instances, it will be necessary to add new stiffener plates and diaphragms at each interior pier. After the splice plates and bearing are in place, the reinforcement required in the deck over the piers is added and a deck replaced. Such a splice is shown in Figure 6.27.

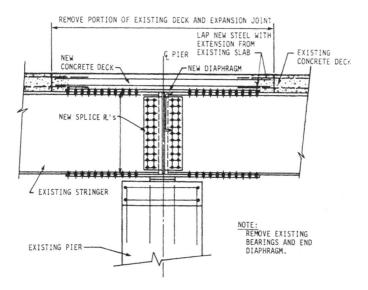

FIGURE 6.27 Conceptual details of a moment- and shear-type connection. (*Source:* Klaiber, F.W. et al., NCHRP 293, Transportation Research Board, 1987. With permission.)

Recently, the Robert Moses Parkway Bridge in Buffalo, New York [78] which originally consisted of 25 simply supported spans ranging from 63 ft (19.2 m) to 77 ft (23.5 m) in length was seismically retrofitted. Moment and shear splices were added to convert the bridge to continuous spans: one two-span element, one three-span element, and five four-span elements. This modification not only strengthened the bridge, but also provided redundancy and improved its earthquake resistance.

Timber Stringers

When providing continuity in timber stringers, steel plates can be placed on both sides and on the top and bottom of the connection and then secured in place with either bolts or lag screws. When adequate plates are used, this provides the necessary moment and shear transfer required. Additional strength can be obtained at the joint by injecting epoxy into the timber cracks as is suggested by Avent et al. [79]. Although adding steel plates requires the design and construction of a detailed connection, significant stress reduction can be obtained through its use.

6.7   Recent Developments

6.7.1   Epoxy Bonded Steel Plates

Epoxy-bonded steel plates have been used to strengthen or repair buildings and bridges in many countries around the world including Australia, South Africa, Switzerland, the United Kingdom, and Japan, to mention a few.

The principle of this strengthening technique is rather simple: an epoxy adhesive is used to bond steel plates to overstressed regions of reinforced concrete members. The steel plates are typically located in the tension zone of a beam; however, plates located in the compression and shear zones have also been utilized. The adhesive provides a shear connection between the reinforced concrete beam and the steel plate, resulting in a composite structural member. The addition of plates in the tension zone not only increases the area of tension steel, but also lowers the neutral axis, resulting in a reduction of live-load stresses in the existing reinforcement. The tension plates effectively increase the flexural stiffness, thereby reducing cracking and deflection of the member.

Although this procedure has been used on dozens of bridges in other countries, to the authors' knowledge, it has not been used on any bridges in the United States due to concerns with the method. Some of these concerns are plate corrosion, long-term durability of the bond connection, plate peeling, and difficulties in handling and installing heavy plates.

In recent years, the steel plates used in this strengthening procedure have been replaced with fiber-reinforced plastic sheets; the most interest has been in carbon fiber-reinforced polymer (CFRP) strips. Although CFRP strips have been used to strengthen various types of structures in Europe and Japan for several years, in the United States there have only been laboratory investigations and some field demonstrations. Discussion in the following section is limited to the use of CFRP in plate strengthening. For information on the use of FRP for increasing the shear strength and ductility of reinforced concrete columns in seismic area, the reader is referred to Reference [91]. This reference is a comprehensive literature review of the various methods of seismic strengthening of reinforced columns.

6.7.2   CFRP Plate Strengthening

CFRP strips have essentially replaced steel plates as CFRP has none of the previously noted disadvantages of steel plates. Although CFRP strips are expensive, the procedure has many advantages: less weight, strengthening can be added to the exact location where increased strength is required, strengthening system takes minimal space, material has high tensile strength, no corrosion problems, easy to handle and install, and excellent fatigue properties. As research is still in progress in Europe, Japan, Canada, and the United States on this strengthening procedure, and since the application of CFRP strips obviously varies from structure to structure, rather than providing details on this procedure, several examples of its application will be described in the following paragraphs.

In 1994, legal truck loads in Japan were increased by 25% to 25 tons. After a review of several concrete slab bridges, it was determined that they were inadequate for this increased load. Approximately 50 of these bridges were strengthened using CFRP sheets bonded to the tension face. The additional material not only reduced the stress in the reinforcing bars, it also reduced the deflections in the slabs due to the high modulus of elasticity of the CFRP sheets.

Recently, a prestressed concrete (P/C) beam in West Palm Beach, Florida, which had been damaged by being struck by an overheight vehicle, was repaired using CFRP. This repair was accomplished in 15 hours by working three consecutive nights with minimal disruption of traffic. The alternative to this repair technique was to replace the damaged P/C with a new P/C beam. This procedure would have taken close to 1 month, and would have required some road closures.

The Oberriet–Meiningen three-span continuous bridge was completed in 1963. This bridge over the Rhine River connects Switzerland and Austria. Due to increased traffic loading, it was determined that the bridge needed strengthening. Strengthening was accomplished in 1996 by increasing the deck thickness 3.1 in. (8 cm) and adding 160 CFRP strips 13.1 ft (4 m) long on 29.5-in. (75-cm) intervals to the underside of the deck. The combination of these two remedies increased the capacity of the bridge so that it is in full compliance with today's safety and load requirements.

Three severely deteriorated 70-year-old reinforced concrete frame bridges near Dreselou, Germany, have recently been strengthened (increased flexure and shear capacity) using CFRP plates. Prior to strengthening, the bridges were restricted to 2-ton vehicles. With strengthening, 16-ton vehicles are now permitted to use the bridges. Prior to implementing the CFRP strengthening procedure, laboratory tests were completed on this strengthening technique at the Technical University in Brauwschweigs, Germany.

6.8   Summary

The purpose of this chapter is to identify and evaluate the various methods of strengthening existing highway bridges and to a lesser extent railroad bridges. Although very few references have been made to railroad bridges, the majority of the strengthening procedures presented could in most situations be applied to railroad bridges.

In this chapter, information on five strengthening procedures (lightweight deck replacement, composite action, strengthening of various bridge members, post-tensioning, and development of bridge continuity) have been presented. A brief introduction to using CFRP strips in strengthening has also been included.

In numerous situations, strengthening a given bridge, rather than replacing it or posting it, is a viable economical alternative which should be given serious consideration.

For additional information on bridge strengthening/rehabilitation, the reader is referred to References [1, 2, 60] which have 208, 379, and 199 references, respectively, on the subject.

Acknowledgments

This chapter was primarily based on NCHRP 12-28(4) "Methods of Strengthening Existing Highway Bridges" (NCHRP Report 293 [2]) which the authors and their colleagues, Profs. K. F. Dunker and W. W. Sanders, completed several years ago. Information from this investigation was supplemented with information from NCHRP Synthesis 249 "Methods for Increasing Live Load Capacity of Existing Highway Bridges" [1], literature reviews, and the results of research projects the authors have completed since submitting the final report to NCHRP 12-28(4).

We wish to gratefully acknowledge the Transportation Research Board, National Research Council, Washington, D.C., who gave us permission to use material from NCHRP 293 and NCHRP Synthesis 249. Information, opinions, and recommendations are from the authors. The Transportation Research Board, the American Association of State Highway and Transportation Officials, and the Federal Highway Administration do not necessarily endorse any particular products, methods, or procedures presented in this chapter.

References

1. Dorton, R. A. and Reel, R., Methods of Increasing Live Load Capacity of Existing Highway Bridges, Synthesis of Highway Practice 249, National Academy Press, Washington, D.C., 1997, 66 pp.
2. Klaiber, F. W., Dunker, K. F., Wipf, T. J., and Sanders, W. W., Jr., Methods of Strengthening Existing Highway Bridges, NCHRP 293, Transportation Research Board, 1987, 114 pp.
3. Sprinkle, M. M., Prefabricated Bridge Elements and Systems, NCHRP Synthesis of Highway Practice 119, Aug. 1985, 75 pp.
4. Shanafelt, G. O. and Horn, W. B., Guidelines for Evaluation and Repair of Prestressed Concrete Bridge Members, NCHRP Report 280, Dec. 1985, 84 pp.
5. Shanafelt, G. O. and Horn, W. B., Guidelines for Evaluation and Repair of Damaged Steel Bridge Members, NCHRP Report 271, June 1984, 64 pp.
6. Applied Technology Council (ATC), Seismic Retrofitting Guidelines for Highway Bridges, Federal Highway Administration, Final Report, Report No. FHWA-RD-83-007, Dec. 1983, 219 pp.
7. Sabnis, G. M., Innovative Methods of Upgrading Structurally and Geometrically Deficient through Truss Bridges, FHWA, Report No. FHWA-RD-82-041, Apr. 1983, 130 pp.
8. University of Virginia Civil Engineering Department, Virginia Highway and Transportation Research Council, and Virginia Department of Highways and Transportation, Bridges on Secondary Highways and Local Roads — Rehabilitation and Replacement, NCHRP Report 222, May 1980, 73 pp.
9. Mishler, H. W. and Leis, B. N., Evaluation of Repair Techniques for Damaged Steel Bridge Members: Phase I, Final Report, NCHRP Project 12-17, May 1981, 131 pp.
10. Shanafelt, G. O. and Horn, W. B., Damage Evaluation and Repair Methods for Prestressed Concrete Bridge Members, NCHRP Report 226, Nov. 1980, 66 pp.
11. University of Virginia Civil Engineering Department, Virginia Highway and Transportation Research Council, and Virginia Department of Highways and Transportation, Bridges on Secondary Highways and Local Roads — Rehabilitation and Replacement, NCHRP Report 222, May 1980, 73 pp.
12. Berger, R. H., Extending the Service Life of Existing Bridges by Increasing Their Load Carrying Capacity, FHWA Report No. FHWA-RD-78-133, June 1978, 75 pp.
13. Fisher, J. W., Hausamann, H., Sullivan, M. D., and Pense, A. W., Detection and Repair of Fatigue Damage in Welded Highway Bridges, NCHRP Report 206, June 1979, 85 pp.
14. Lichenstein, A. G., Bridge Rating through Nondestructive Load Testing, NCHRP Project 12-28(13)A, 1993, 117 pp.
15. Timmer, D. H., A study of the concrete filled steel grid bridge decks in Ohio, in *Bridge Maintenance and Rehabilitation Conference,* Morgantown, WV, Aug. 13–16, 1980, 422–475.
16. DePhillips, F. C., Bridge deck installed in record time, *Public Works,* 116(1), 76–77, 1985.
17. Bakht, B. and Tharmabala, T., Steel-wood composite bridges and their static load performance, in *Can. Soc. Civ. Eng. Annual Conference,* Saskatoon, Canada, May 27–31, 1985, 99–118.
18. Csagoly, P. F. and Taylor, R. J., A structural wood system for highway bridges, in *International Association for Bridge and Structural Engineering 11th Congress,* Final Report, Italy, Aug. 31–Sept. 5, 1980, 219–225.
19. Taylor, R. J., Batchelor, B. D., and Vandalen, K., Prestressed wood bridges, *Proc. International Conference on Short and Medium Span Bridges,* Toronto, Aug. 8–12, 1982, 203–218.
20. Ministry of Transportation and Communications, Design of Wood Bridge Using the Ontario Highway Bridge Design Code, Ontario, Canada, 1983, 72 pp.
21. Ministry of Transportation and Communications, Design of Prestressed Wood Bridges Using the Ontario Highway Bridge Design Code, Ontario, Canada, 1983, 30 pp.
22. Muchmore, F. W., Techniques to bring new life to timber bridges, *ASCE J. Struct. Eng.,* 110(8), 1832–1846, 1984.

23. Mackie, G. K., Recent uses of structural lightweight concrete, *Concrete Constr.*, 30(6), 497–502, 1985.
24. Steel grids rejuvenate old bridges, in *The Construction Advisor,* Associated General Contractors of Missouri, Jefferson City, MO, May 1974, 18–19.
25. Lightweight decking rehabs downrated Ohio River span, *Rural Urban Roads*, 20(4), 25–26, 1982.
26. CAWV Members Join Forces to Reinforce Bridge, Reprint from *West Virginia Construction News,* 1982, 3 pp.
27. The rehabilitation of the Old York Road Bridge, *Rural Urban Roads,* 21(4), 22–23, 1983.
28. Campisi, V. N., Exodermic deck systems: a recent development in replacement bridge decks, *Modern Steel Constr.*, 26(3), 28–30, 1986.
29. Holm, T. A., Structural lightweight concrete for bridge redecking, *Concrete Constr.*, 30(8), 667–672, 1985.
30. Ford, J. H., Use of Precast, Prestressed Concrete for Bridge Decks, Progress Report, Purdue University and Indiana State Highway Commission, Joint Highway Research Project No. C-36-56N, No. 20, July 1969, 161 pp.
31. Kropp, P. K., Milinski, E. L., Gutzwiller, M. J., and Lee, R. H., Use of Precast-Prestressed Concrete for Bridge Decks, Final Report, Purdue University and Indiana State Highway Commission, Joint Highway Research Project No. C-36-56N, July 1975, 85 pp.
32. Greiner Engineering Sciences, Inc., Widening and replacement of concrete deck of Woodrow Wilson Memorial Bridge, Paper presented at Session 187 (Modular Bridge Decks), 62nd Annual Transportation Research Board Meeting, Jan. 20, 1983, 16 pp.
33. Nickerson, R. L., Bridge rehabilitation–construction view expediting bridge redecking, in *Proc. 2nd Annual International Bridge Conference,* Pittsburgh, PA, June 17–19, 1985, 5–9.
34. Stemler, J. R., Aluminum orthotropic bridge deck system, Paper presented at 3rd Annual International Bridge Conference, Pittsburgh, PA, June 2–4, 1986, 62–65.
35. Woeltinger, O. and Bock, F., The alteration of the High Bridge Levensau over the North-East Canal [Der Umbau der Hochbruecke Levensau über den Nord-Ostsee-Kanal], *Stahlbau,* West Germany, 23(12), 295–303, 1956 [in German].
36. Freudenberg, G., Raising the load capacity of an old bridge, [Erhoehung der Tragfaehigkeit einer alten Bruecke], *Stahlau,* West Germany, 48(3), 76–78, 1979 [in German].
37. Bridge rebuilt with composite design, *Eng. News Rec.,* 165(20), 91, 1960.
38. Collabella, D., Rehabilitation of the Williamsburg and Queensboro Bridges — New York City, *Municipal Eng. J.,* 70 (Summer), 1984, 30 pp.
39. Cook, J. P., The shear connector, *Composite Construction Methods,* John Wiley & Sons, New York, 1977, 168–172.
40. Klaiber, F. W., Dedic, D. J., Dunker, K. F., and Sanders, W. W., Jr., Strengthening of Existing Single Span Steel Beam and Concrete Deck Bridges, Final Report — Part I, Engineering Research Institute Project 1536, ISU-ERI-Ames-83185, Iowa State University, Feb. 1983, 185 pp.
41. Dedic, D. J. and Klaiber, F. W., High strength bolts as shear connectors in rehabilitation work, *Concrete Int. Des. Constr.* 6(7), 41–46, 1984.
42. Dallam, L. N., Pushout Tests with High-Strength Bolt Shear Connectors, Missouri Cooperative Highway Research Program Report 68-7, Engineering Experiment Station, University of Missouri-Columbia, 1968, 66 pp.
43. Dallam, L. N., Static and Fatigue Properties of High-Strength Bolt Shear Connectors, Missouri Cooperative Highway Research Program Report 70-2, Engineering Experiment Station, University of Missouri-Columbia, 1970, 49 pp.
44. Dunker, K. F., Klaiber, F. W., Beck, B. L., and Sanders, W. W., Jr., Strengthening of Existing Single-Span Steel-Beam and Concrete Deck Bridges, Final Report — Part II, Engineering Research Institute Project 1536, ISU-ERI-Ames-85231, Iowa State University, March 1985, 146 pp.
45. Watter, F., Albrecht, P., and Sahli, A. H., End bolted cover plates, *ASCE J. Struct. Eng.,* 111(6), 1235–1249, 1985.

46. Park, S. H., *Bridge Rehabilitation and Replacement (Bridge Repair Practice)*, S. H. Park, Trenton, NJ, 1984, 818 pp.

47. Albrecht, P., Watter, F., and Sahli, A., Toward fatigue-proofing cover plate ends, in *Proc. W. H. Munse Symposium on Behavior of Metal Structures, Research to Practice*, ASCE National Convention, Philadelphia, PA, May 17, 1983, 24–44.

48. Taylor, R., Strengthening of reinforced and prestressed beams, *Concrete*, 10(12), 28–29, 1976.

49. Warner, R. F., Strengthening, stiffening and repair of concrete structures, *Int. Assoc. Bridge Struct. Eng. Surv.*, 17 May, 25–41, 1981.

50. Westerberg, B., Strengthening and repair of concrete structures [Forstarkning och reparation av betongkonstruktioner], *Nord. Betong*, Sweden, 7–13, 1980 [in Swedish].

51. Mancarti, G. D., Resurfacing, restoring and rehabilitating bridges in California, in *Proc. International Conference on Short and Medium Span Bridges*, Toronto, Aug. 8–12, 1982, 344–355.

52. Rodriguez, M., Giron H., and Zundelevich, S., Inspection and design for the rehabilitation of bridges for Mexican railroads, *Proc. 2nd Annual International Bridge Conference*, Pittsburgh, PA, June 17–19, 1985, 12 pp.

53. Warner, R. F., Strengthening, stiffening and repair of concrete structures, in *Proc. International Symposium on Rehabilitation of Structures*, Maharastra Chapter of the American Concrete Institute, Bombay, Dec. 1981, 187–197.

54. OECD, Organization for Economic Co-Operation and Development Scientific Expert Group, *Bridge Rehabilitation and Strengthening*, Paris, 1983, 103 pp.

55. Suter, R. and Andrey, D., Rehabilitation of bridges [Assainissement de ponts] *Inst. Statique et Structures Beton Arme et Precontraint*, Switzerland, 106 (Mar.) 105–115, 1985 [in French].

56. Dilger, W. H. and Ghali, A., Remedial measures for cracked webs of prestressed concrete bridges, *J. PCI*, 19(4), 76–85, 1984.

57. West, J. D., Some methods of extending the life of bridges by major repair or strengthening, *Proc. ICE*, 6 (Session 1956–57), 183–215, 1957.

58. Stratton, F. W., Alexander, R., and Nolting, W., Development and Implementation of Concrete Girder Repair by Post-Reinforcement, Kansas Department of Transportation, May 1982, 31 pp.

59. Soliman, M. I., Repair of distressed reinforced concrete columns, in *Can. Soc. Civ. Eng. Annual Conference*, Saskatoon, Canada, May 27–31, 1985, 59–78.

60. Wipf, T. J., Klaiber, F. W., and Russo, F. M., Evaluation of Seismic Retrofit Methods for Reinforced Concrete Bridge Columns, Technical Report NCEER-97-0016, National Center for Earthquake Engineering Research, Buffalo, NY, Dec. 1997, 168 pp.

61. Karamchandani, K. C., Strengthening of Reinforced Concrete Members, in *Proc. International Symposium on Rehabilitation of Structures*, Maharastra Chapter of the American Concrete Institute, Bombay, Dec. 1981, 157–159.

62. Klein, G. J. and Gouwens, A. J., Repair of columns using collars with circular reinforcement, *Concrete Int. Des. Constr.*, 6(7) 23–31, 1984.

63. Rheinisches Strassenbauamt Moenchengladbach, Rehabilitation of Structure 41 in Autobahnkreuz Holz [Erfahrungsbericht–Sanierung des Bauwerks Nr. 41 in Autobahnkreuz Holz], Rheinisches Strassenbauamt Moenchengladbach, Germany, 1983, 11 pp. [in German].

64. Lee, D. H., Prestressed concrete bridges and other structures, *Struct. Eng.*, 30(12), 302–313, 1952.

65. Jungwirth, D. and Kern, G., Long-term maintenance of prestressed concrete structures — prevention, detection and elimination of defects [Langzeitverhalten von Spannbeton — Konstruktionen Verhueten, Erkennen und Beheben von Schaden], *Beton-Stahlbetonbau*, West Germany, 75(11), 262–269, 1980 [in German].

66. Mancarti, G. D., Strengthening California's Steel Bridges by Prestressing, TRB Record 950, Transportation Research Board, 1984, 183–187.

67. Vernigora, E., Marcil, J. R. M., Slater, W. M., and Aiken, R. V., Bridge rehabilitation and strengthening by continuous post-tensioning, *J. PCI*, 14(2) 88–104, 1969.

68. Ferjencik, P. and Tochacek, M., *Prestressing in Steel Structures* [Die Vorspannung in Stahlbau], Wilhelm Ernst & Sohn, West Germany, 1975, 406 [in German].

69. Kandall, C., Increasing the load-carrying capacity of existing steel structures, *Civ. Eng.*, 38(10), 48–51, 1968.

70. Benthin, K., Strengthening of Bridge No. 3699, Chaska, Minnesota, Minnesota Department of Transportation, 1975, 11 pp.

71. Andrey, D. and Suter, R., *Maintenance and Repair of Construction Works* [Maintenance et reparation de ouvrages d'art], Ecole Polytechnique Federale de Lausanne, Lausanne, Switzerland, 1986 [in French].

72. Mueller, T., Alteration of the highway bridge over the Aare River in Aarwangen [Umbau der Strassenbruecke über die Aare in Aarwangen], *Schweiz. Bauz.*, 87(11), 199–203, 1969 [in German].

73. Taylor, R. J. and Walsh, H., Prototype Prestressed Wood Bridge, *TRB Report 950*, Transportation Research Board, 1984, 110–122.

74. Lamberson, E. A., Post-Tensioning Concepts for Strengthening and Rehabilitation of Bridges and Special Structures: Three Case Histories of Contractor Initiated Bridge Redesigns, Dywidag Systems International, Lincoln Park, NJ, 1983, 48 pp.

75. Dunker, K. F., Klaiber, F. W., and Sanders, W. W., Jr., Post-tensioning distribution in composite bridges, *J. Struct. Eng. ASCE*, 112 (ST11), 2540–2553, 1986.

76. El-Arabaty, H. A., Klaiber, F. W., Fanous, F. S., and Wipf, T. J., Design methodology for strengthening of continuous-span composite bridges, *J. Bridge Eng. ASCE*, 1(3), 104–111, 1996.

77. Roberts, J., Manual for Bridge Maintenance Planning and Repair Methods, Florida Department of Transportation, 1978, 282 pp.

78. Malik, A. H., Seismic Retrofit of the Robert Moses Parkway Bridge, in *Proc. 12th U.S.–Japan Bridge Engineering Workshop*, Buffalo, NY, Oct. 1996, 215–228.

79. Avent, R. R., Emkin, L. Z., Howard, R. H., and Chapman, C. L., Epoxy-repaired bolted timber connections, *J. Struct. Div. Proc. ASCE*, 102(ST4), 821–838, 1976.

7

Cable Force Adjustment and Construction Control

Danjian Han
*South China University
of Technology*

Quansheng Yan
*South China University
of Technology*

7.1 Introduction

Due to their aesthetic appeal and economic advantages, many cable-stayed bridges have been built over the world in the last half century. With the advent of high-strength materials for use in the cables and the development of digital computers for the structural analysis and the cantilever construction method, great progress has been made in cable-stayed bridges [1,2]. The Yangpu Bridge in China with a main span of 602 m, completed in 1993, is the longest cable-stayed bridge with a composite deck. The Normandy Bridge in France, completed in 1994, with main span of 856 m is now the second-longest-span cable-stayed bridge. The Tatara Bridge in Japan, with a main span of 890 m, was opened to traffic in 1999. More cable-stayed bridges with larger spans are now in the planning.

Cable-stayed bridges are featured for their ability to have their behavior adjusted by cable stay forces [3–5]. Through the adjustment of the cable forces, the internal force distribution can be optimized to a state where the girder and the towers are compressed with little bending. Thus, the performance of material used for deck and pylons can be efficiently utilized.

During the construction of a cable-stayed bridge there are two kinds of errors encountered frequently [6,13]: one is the tension force error in the jacking cables, and the other is the geometric error in controlling the elevation of the deck. During construction the structure must be monitored and adjusted; otherwise errors may accumulate, the structural performance may be substantially influenced, or safety concerns may arise. With the widespread use of innovative construction methods, construction control systems play a more and more important role in construction of cable-stayed bridges [18,19].

There are two ways of adjustment: adjustment of the cable forces and adjustment of the girder elevations [7]. The cable-force adjustment may change both the internal forces and the configuration of the structure, while the elevation adjustment only changes the length of the cable and does not induce any change in the internal forces of the structure.

This chapter deals with two topics: cable force adjustment and construction control. The methods for determing the cable forces are discussed in Section 7.2, then a presentation of the cable force adjustment is given in Section 7.3. A simulation method for a construction process of prestressed concrete (PC) cable-stayed bridge is illustrated in Section 7.4, and a construction control system is introduced in Section 7.5.

7.2   Determination of Designed Cable Forces

For a cable-stayed bridge the permanent state of stress in a structure subjected to dead load is determined by the tension forces in the cable stays. The cable tension can be chosen so that bending moments in the girders and pylons are eliminated or at least reduced as much as possible. Thus the deck and pylon would be mainly under compression under the dead loads [3,10].

In the construction period the segment of deck is corbeled by cable stays and each cable placed supports approximately the weight of one segment, with the length corresponding to the longitudinal distance between the two stays. In the final state the effects of other dead loads such as wearing surface, curbs, fence, etc., as well as the traffic loads, must also be taken into account. For a PC cable-stayed bridge, the long-term effects of concrete creep and shrinkage must also be considered [4].

There are different methods of determining the cable forces and these are introduced and discussed in the following.

7.2.1   Simply Supported Beam Method

Assuming that each stayed cable supports approximately the weight of one segment, corresponding to the longitudinal distance between two stays, the cable forces can be estimated conveniently [3,4]. It is necessary to take into account the application of other loads (wearing surface, curbs, fences, etc.). Also, the cable is placed in such a way that the new girder element is positioned correctly, with a view to having the required profile when construction is finished.

Due to its simplicity and easy hand calculation, the method of the simply supported beam is usually used by designers in the tender and preliminary design stage to estimate the cable forces and the area of the stays. For a cable-stayed bridge with an asymmetric arrangement of the main span and side span or for the case that there are anchorage parts at its end, the cable forces calculated by this method may not be evenly distributed. Large bending moments may occur somewhere along the deck and/or the pylons which may be unfavorable.

7.2.2   Method of Continuous Beam on Rigid Supports

By assuming that under the dead load the main girder behaves like a continuous beam and the inclined stay cables provide rigid supports for the girder, the vertical component of the forces in stay cables are equal to the support reactions calculated on this basis [4,10]. The tension in the anchorage cables make it possible to design the pylons in such a way that they are not subjected to large bending moments when the dead loads are applied.

This method is widely used in the design of cable-stayed bridges. Under the cable forces calculated by this method, the moments in the deck are small and evenly distributed. This is especially favorable for PC cable-stayed bridges because the redistribution of internal force due to the effects of concrete creep could be reduced.

7.2.3 Optimization Method

In the optimization method of determining the stresses of the stay cables under permanent loads, the criteria (objective functions) are chosen so the material used in girders and pylons is minimized [8,11]. When the internal forces, mainly the bending moments, are evenly distributed and small, the quantity of material reaches a minimum value. Also the stresses in the structure and the deflections of the deck are limited to prescribed tolerances.

In a cable-stayed bridge, the shear deformations in the girder and pylons are neglected, the strain energy can be represented by

$$U = \frac{1}{2}\int_0^L \frac{M^2}{2\,EI}dx + \frac{1}{2}\int_0^L \frac{N^2}{2\,EA}dx \tag{7.1}$$

where EI is the bending stiffness of girder and pylons and EA is the axial stiffness.

It can be given in a discrete form when the structure is simulated by a finite-element model as

$$U = \sum_{i=1}^N \frac{L_i}{4\,E_i}\left(\frac{M_{il}^2 + M_{ir}^2}{l_i} + \frac{N_{il}^2 + N_{ir}^2}{A_i}\right) \tag{7.2}$$

where N is the total number of the girder and pylon elements, L_i is the length of the ith element, E is the modulus of elasticity, and I_i and A_i are the moment of inertia and the sections area, respectively. M_{ir}, M_{il}, N_{ir}, and N_{il} are the moments and the normal forces in the left and right end section of the ith element, respectively.

Under the application of dead loads and cable forces the bending moments and normal forces of the deck and pylon are given by

$$\{M\} = \{M_D\} + \{M_P\} = \{M_D\} + [S_M]*\{P_0\} \tag{7.3a}$$

$$\{N\} = \{N_D\} + \{N_P\} = \{N_D\} + [S_N]*\{P_0\} \tag{7.3b}$$

where $\{M_D\}$ and $\{M_P\}$ are the bending moment vectors induced by dead loads and the cable forces, respectively; $[S_M]$ is the moment influence matrix; and $[S_N]$ is the normal force influence matrix, the component S_{ij} of influence matrix represents changes of the moment or the normal force in the ith element induced by the jth unit cable force. $\{N_D\}$ and $\{N_P\}$ are the normal force vectors induced by dead loads and cable forces, respectively. $\{P_0\}$ is the vector of cable forces.

The corresponding displacements in deck and pylon are given as

$$\{F\} = \{F_D\} + \{F_P\} = \{F_D\} + [S_F]*\{P_0\} \tag{7.4}$$

where $\{F\}$ is the displacement vector, $\{S_F\}$ is the displacement influence matrix, and $\{F_D\}$ and $\{F_P\}$ are the displacement vectors induced by dead loads and by cable forces, respectively.

Substitute Eqs. (7.3a) and (7.3b) into Eq. (7.2), and replace the variables by

$$\{\overline{M}\} = [A]\{M\}, \{\overline{N}\} = [B]\{N\} \tag{7.5}$$

in which $[A]$ and $[B]$ are diagonal matrices:

$$[A] = Diag\left[\sqrt{L_1 \Big/ 4E_1 I_1}, \sqrt{L_2 \Big/ 4E_2 I_2}, \cdots, \sqrt{L_n \Big/ 4E_n I_n}\right]$$

$$[B] = Diag\left[\sqrt{L_1 \Big/ 4E_1 A_1}, \sqrt{L_2 \Big/ 4E_2 A_2}, \cdots, \sqrt{L_n \Big/ 4E_n A_n}\right]$$

Then the strain energy of the cable-stayed bridge can be represented in matrix form as

$$U = \{P_0\}^T [\bar{S}]^T [\bar{S}]\{P_0\} + 2\{\bar{P}_D\}^T [\bar{S}]\{P_0\} + \{\bar{P}_D\}^T \{\bar{P}_D\} \tag{7.6}$$

in which $[\bar{S}] = (\bar{S}_M, \bar{S}_N)^T = [A, B](S_M, S_N)^T$, and $\{\bar{P}_D\} = \{M_D, N_D\}^T$.

Now, we want to minimize the strain energy of structure, i.e., to let

$$\partial U \Big/ \partial P_0 = 0 \tag{7.7}$$

under the following constraint conditions:

1. The stress range in girders and pylons must satisfy

$$\{\sigma\}_L \leq \{\sigma\} \leq \{\sigma\}_U \tag{7.8}$$

   in which $\{\sigma\}$ is the maximum stress value vector. And $\{\sigma\}_L, \{\sigma\}_U$ are vectors of the lower and upper bounds.

2. The stresses in stay cables are limited so that the stays can work normally.

$$\{\sigma\}_{LC} \leq \left\{\frac{P_{0C}}{A_C}\right\} \leq \{\sigma\}_{UC} \tag{7.9}$$

   in which A_C is the area of a stay, P_{0C} is the cable force and $\{\sigma\}_{LC}, \{\sigma\}_{UC}$ represent the lower and upper bounds, respectively.

3. The displacements in the deck and pylon satisfy

$$\{|D_i|\} \leq \{\Delta\} \tag{7.10}$$

   in which the left hand side of Eq. (7.10) is the absolute value of maximum displacement vector and the right-hand side is the allowable displacement vector.

Eqs. (7.6) and (7.7) in conjunction with the conditions (7.8) through (7.10) is a standard quadric programming problem with constraint conditions. It can be solved by standard mathematical methods.

Since the cable forces under dead loads determined by the optimization method are equivalent to the cable force under which the redistribution effect in the structure due to concrete creep is minimized [8], the optimization method is used more widely in the design of PC cable-stayed bridges.

7.2.4 Example

For a PC cable-stayed bridge as shown in Figures 7.1 and 7.2, the forces of cable stays under permanent loads (not taking into account the creep and shrinkage) can be determined by the above methods. The results obtained are shown in Figures 7.3 and 7.4. In these figures SB represents the "Simply Supported Beam Method," CB the "Continuous Beam on Rigid Support Method," OPT the "Optimization Method," M the middle span, and S the side span numbering from the pylon location.

As can be seen, because the two ends of the cable-stayed bridge have anchored parts the cable forces located in these two regions obtained by the method of simply supported beam (SB) and by the method of continuous beam on rigid supports (CB) are not evenly distributed. The cable forces in the region near the pylon are very different with the three methods. In the other regions there is no prominent difference among the cable forces obtained by SB, CB, and OPT.

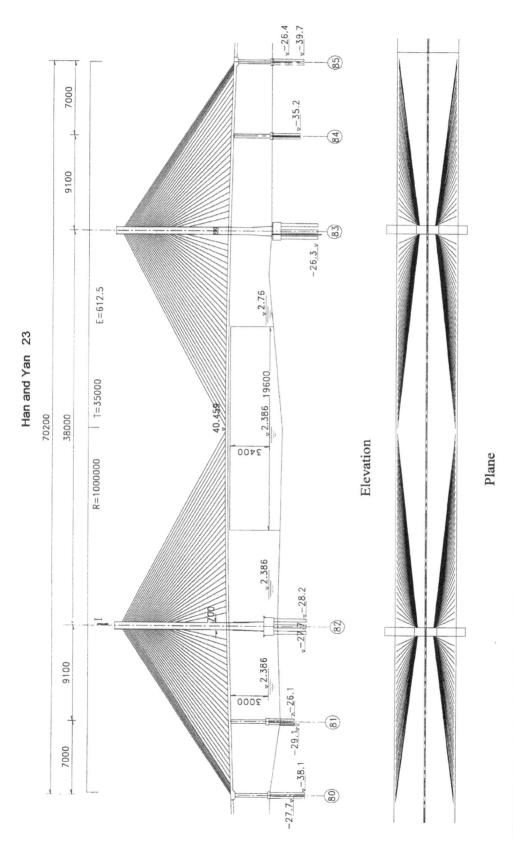

FIGURE 7.1 General view of a PC cable-stayed bridge.

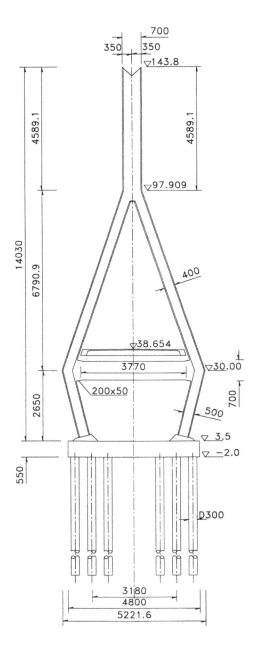

FIGURE 7.2 Side view of tower.

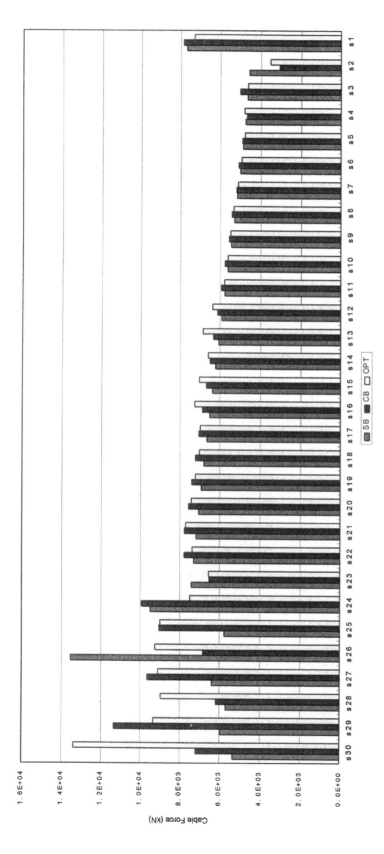

FIGURE 7.3 Comparison of the cable forces (kN) (side span).

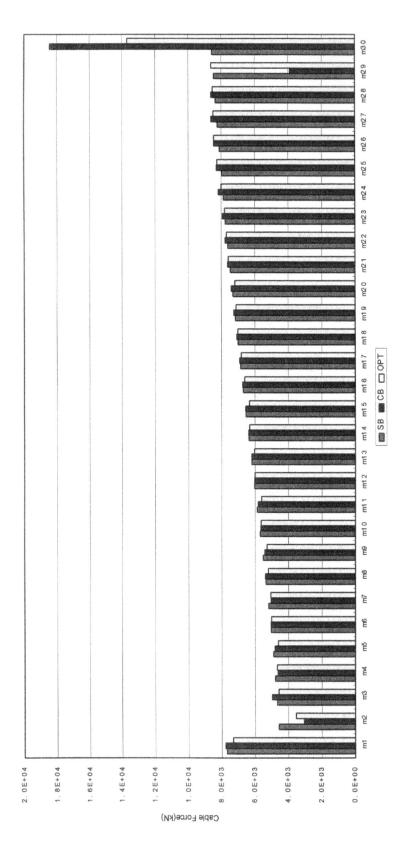

FIGURE 7.4 Comparison of the cable force (middle span).

Generally speaking, the differences of cable forces under dead loads obtained by the above methods are not so significant. The method of simply supported beam is the most convenient and the easiest to use. The method of continuous beam on rigid supports is suitable to use in the design of PC cable-stayed bridges. The optimization method is based on a rigorous mathematical model. In practical engineering applications the choice of the above methods is very much dependent on the design stage and designer preference.

7.3 Adjustment of the Cable Forces

7.3.1 General

During the construction or service stage, many factors may induce errors in the cable forces and elevation of the girder, such as the operational errors in tensioning stays or the errors of elevation in laying forms [14,16,17]. Further, the discrepancies of parameter values between design and reality such as the modules of elasticity, the mass density of concrete, the weight of girder segments may give rise to disagreements between the real structural response and the theoretical prediction [13]. If the structure is not adjusted to reduce the errors during construction, they may accumulate and the structure may deviate away from the intended design aim. Moreover, if the errors are greater than the allowable limits, they may give rise to the unfavorable effects to the structure. Through cable force adjustment, the construction errors can be eliminated or reduced to an allowable tolerance. In the service stage, because of concrete creep effects, cable force may need to be adjusted; thus, an optimal structural state can be reached or recovered.

7.3.2 Influence Matrix of the Cable Forces

Assuming that a unit amount of cable force is adjusted in one cable stay, the deformations and internal forces of the structure can be calculated by finite-element model. The vectors of change in deformations and internal forces are defined as influence vectors. In this way, the influence matrices can be formed for all the stay cables.

7.3.3 Linear Programming Method [7]

Assume that there are n cable stays whose cable forces are going to be adjusted, the adjustments are T_i $(i = 1,2,...,n)$, these values form a vector of cable force adjustment $\{T\}$ as

$$\{T\} = \{T_1, T_2, ..., T_n\}^T \tag{7.11}$$

Denote internal force influence vector $\{P_l\}$ as

$$\{P_l\} = \left(P_{l1}, P_{l2}, ..., P_{1n}\right)^T \quad (l = 1,2,...,m) \tag{7.12}$$

in which m is the number of sections of interest and P_{ij} is the internal force increment due to a unit tension of the jth cable. Denote displacement influence vector $\{D_i\}$ as

$$\{D_i\} = \left(D_{i1}, D_{i2}, ..., D_{in}\right)^T \quad (i = 1,2,...,k) \tag{7.13}$$

in which k is the number of sections of interest and D_{ij} is the displacement increment at section i due to a unit tension of the jth cable. Thus, the influence matrices of internal forces and displacements are given by

$$\{P\} = \left(P_1, P_2, ..., P_m\right)^T \tag{7.14a}$$

$$\{D\} = \left(D_1, D_2, ..., D_k\right)^T \tag{7.14b}$$

respectively. Under application of the cable force adjustment $\{T\}$ (see Eq. (7.11)), the increment of the internal forces and displacements can be given by

$$\{\Delta P\} = \{P\}^T \{T\} \tag{7.15a}$$

$$\{\Delta D\} = \{D\}^T \{T\} \tag{7.15b}$$

respectively.

Denote deflection error vector $\{H\}$ as

$$\{H\} = \left(h_1, h_2, \ldots\ldots, h_m\right)^T \tag{7.16}$$

Denote vector of internal force $\{N\}$ as

$$\{N\} = \left(N_1, N_2, \ldots\ldots, N_m\right)^T \tag{7.17}$$

After cable force adjustments, the absolute values of the deflection errors are expressed by

$$\left|\lambda_k\right| = \left|\sum_{i=1}^{n} D_{ik} T_i - h_k\right| \tag{7.18}$$

and the absolute values of internal force errors are expressed by

$$\left|q_l\right| = \left|\sum_{i=1}^{n} P_{il} T_i - N_l\right| \tag{7.19}$$

The objective function for cable force adjustments may be defined as the errors of girder elevation, i.e.,

$$\min\left|\lambda_k\right| \tag{7.20}$$

and the constraint conditions may include limitations of the internal force errors, the upper and lower bounds of the cable forces, and the maximum stresses in girders and pylons. Then the optimum values of cable force adjustment can be determined by a linear programming model.

The value of cable adjustment $\{T\}$ could be positive for increasing or negative for decreasing of the cable forces. Introduce two auxiliary variables T_{1i}, T_{2i} as

$$T_i = T_{1i} - T_{2i} \quad T_{1i} \geq 0, T_{2i} \geq 0 \tag{7.21}$$

Substitute Eq. (7.18) into (7.20), then a linear program model is established by

$$\min: \lambda_k \tag{7.22}$$

subject to

$$\sum_{i=1}^{n} p_{li}(T_{1i} - T_{2i}) \geq N_l - \xi \bar{p}_l \ (l = 1,2,\ldots m) \tag{7.23a}$$

$$\sum_{i=1}^{n} p_{li}(T_{1i} - T_{2i}) \leq N_l + \xi \bar{p}_l \ (l = 1,2,\ldots m) \tag{7.23b}$$

$$\sum_{i=1}^{n} D_{ji}(T_{1i} - T_{2i}) \geq h_j - d_j \ (j = 1,2,\dots k) \tag{7.23c}$$

$$\sum_{i=1}^{n} D_{ji}(T_{1i} - T_{2i}) \geq h_j - d_j \ (j = 1,2,\dots k) \tag{7.23d}$$

$$T_{1i} - T_{2i} \leq \eta \overline{T_i}, \quad T_{1i} - T_{2i} \geq -\eta \overline{T_i} \ (I = 1,2,\dots,n) \tag{7.23e}$$

in which $\overline{p_l}$ is the design value of internal force at section l, ξ is the allowable tolerance in percentage of the internal force, $\overline{T_i}$ is the design value of the cable force, and η is the allowable tolerance in percentage of the cable forces.

Equations (7.22) and (7.23) form a standard linear programming problem that can be solved by mathematical software.

7.3.4 Order of Cable Adjustment

The adjustment values can be determined by the above method; however, the adjustments must be applied at the same time to all cables, and a great number of jacks and workers are needed [7]. In performing the adjustment, it is preferred that the cable stays are tensioned one by one.

When adjusting the cable force individually, the influence of the other cable forces must be considered. And since any cable must be adjusted only one time, the adjustment values can be calculated through the influence matrix of cable force.

$$\{\overline{T}\} = [S]\{T\} \tag{7.24}$$

where $\{\overline{T}\} = \{\overline{T_1}, \overline{T_2}, \dots \overline{T_n}\}$ is the vector of actual adjustment value of cable tension. $[S]$ is the influence matrix of cable tension, whose component S_{ij} represents tension change of the jth cable when the ith cable changes a unit amount of force.

7.4 Simulation of Construction Process

7.4.1 Introduction

Segmental construction techniques have been widely used in construction of cable-stayed bridges. In this technique, the pylon(s) is built first; then the girder segments are erected one by one and supported by the inclined cables hung from the pylon(s). It is evident that the profile of the main girder and the final tension forces in the cables are strongly related to the erection method and the construction scheme. It is therefore important that the designer should be aware of the construction process and the necessity to look into the structural performance at every stage of construction [9,12].

In any case, structural safety is the most important issue since the stresses in the girder and pylon(s) are related to the cable tensions. Thus the cable forces are of great concern. Further, during construction, the geometric profile of the girder is also very important. It is clear that if the profile of the girder were not smooth or, finally, the cantilever ends could not meet together, then the construction might experience some trouble. The profile of the girder or the elevation of the bridge segments is mainly controlled by the cable lengths. Therefore, the cable length must be appropriately set at the erection of each segment. It also should be noted that in the construction process, the internal forces of the structure and the elevation of the girder could vary because usually the bridge segments are built by a few components at a time and the erection equipment is placed at different positions during construction and because some errors such as the weight of the segment and the tension force of the cable, etc., may occur. Thus, monitoring and adjustment are absolutely needed.

To reach the design aim, an effective and efficient simulation of the construction process step by step is very necessary. The objectives of the simulation analysis are [4,12]:

1. To determine the forces required in cable stays at each construction stage;
2. To set the elevation of the girder segment;
3. To find the consequent deformation of the structure at each construction stage;
4. To check the stresses in the girder and pylon sections.

The simulation methods are introduced and discussed in detail in the following sections. In the next section, the technique of forward analysis is presented to simulate the assemblage process. Creep effects can be considered; however, the design aim may not be successfully achieved by such simulation because it is not so easy to determine the appropriate lengths of the cable stays which make the final elevation to achieve the design profile automatically. Another technique presented is the backward disassemblage analysis, which starts with the final aim of the structural state and disassembles segment by segment in a reverse way. The disadvantage of this method is that the creep effects may not be able to be defined. However, values obtained from the assemblage process may be used in this analysis. These two methods may be alternatively applied until convergence is reached.

It is noted that the simulation is only limited to that of the erection of the superstructure.

7.4.2 Forward Assemblage Analysis

Following the known erection procedure, a simulation analysis can be carried out by the finite-element method. This is the so-called forward assemblage analysis. It has been used to simulate the erection process for PC bridges built by the free cantilever method.

Concerning finite-element modeling, the structure may be treated as a plane frame or a space frame [4]. A plane frame model may be good enough for construction simulation because transverse loads, such as wind, can generally be ignored. In a plane frame model, the pylon(s) and the girder are modeled by some beam elements, while the stays are modeled as two-node bar elements with Ernst modules [3,4] by which the effects of cable sag can be taken into account. The structural configuration is changed stage by stage. Typically, in one assemblage stage, a girder segment treated as one or several beam elements is connected to the existing structure, while its weight is treated as a load to apply to the element. Also, the cable force is applied. Then an analysis is performed and the structure is changed to a new configuration.

In finite-element modeling, several factors such as the construction loads (weight of equipment and traveling carriage) and effects of concrete creep and shrinkage, must be considered in detail.

Traveling carriages are specially designed for construction of a particular bridge project. Generally there are two kinds of carriages. One is cantilever type (Figure 7.5a). The traveling carriages are mounted near the ends of girders, like a cantilever to support the next girder segment. In this case, the weight of the carriage can be treated as an external load applied to the end of the girder.

With the development of multiple cable systems, the girder with lower height becomes more flexible. The girder itself is not able to carry the cantilever weights of the carriage and the segment. Then an innovative erection technique was proposed [1]. And another type of carriage is developed. This new idea is to use permanent stays to support the form traveler (Figure 7.5b) so that the concrete can be poured *in situ* [1,9]. This method enjoys considerable success at present because of the undeniable economic advantages. Its effectiveness has been demonstrated by bridge practice. For the erecting method using the later type of carriage, the carriage works as a part of the whole structure when the segmental girder is poured *in situ*. Thus, the form traveler must be included in the finite-element model to simulate construction. A typical flowchart of forward assemblage analysis is shown in Figure 7.6.

With the forward assemblage analysis, the construction data can be worked out. And the actual permanent state of cable-stayed bridges can be reached. Further, if the erection scheme were modified during the construction period or in the case that significant construction error occurred, then the structural parameters or the temporary erection loads would be different from the values used in the design. It is possible to predict the cable forces and the sequential deformations at each stage by utilizing the forward assemblage analysis.

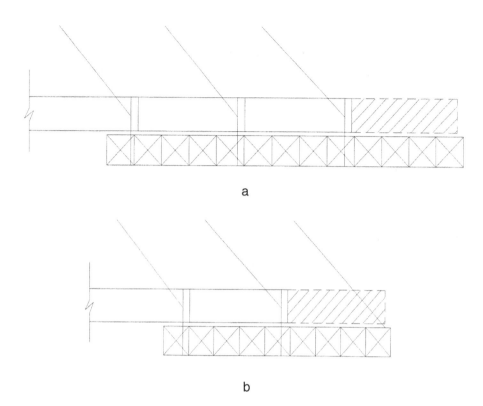

a

b

FIGURE 7.5   (a) Cantilever carriage; (b) cable-supported cantilever carriage.

7.4.3   Backward Disassemblage Analysis

Following a reverse way to simulate the disassemblage process stage by stage, a backward analysis can be carried out also by finite-element method [11,12]. Not only the elevations of deck but also the length of cable stays and the initial tension of stays can be worked out by this method. And the completed state of structure at each stage can be evaluated.

The backward disassemblage analysis starts with a very ideal structural state in which it is assumed that all the creep and shrinkage deformation of concrete be completed, i.e., a state 5 years or 1500 days after the completion of the bridge construction. The structural deformations and internal forces at each stage are considered ideal reference states for construction of the bridge [11]. The backward analysis procedure for a PC cable-stayed bridge may be illustrated as follows:

    Step 1. Compute the permanent state of the structure.

    Step 2. Remove the effects of the creep and shrinkage of concrete of 1500 days or 5 years.

    Step 3. Remove the second part of the dead loads, i.e., the weights of wearing surfacing, curbs and
           fence, etc.

    Step 4. Apply the traveler and other temporary loads and supports.

    Step 5. Remove the center segment, to analyze the semistructure separately.

    Step 6. Move the form traveler backward.

    Step 7. Remove the weight of the concrete of a pair of segments.

    Step 8. Remove the cable stay.

    Step 9. Remove the corresponding elements.

Repeat the Steps 6 to 9 until all the girder segments are disassembled. A flowchart for backward disassemblage analysis is shown in Figure 7.7.

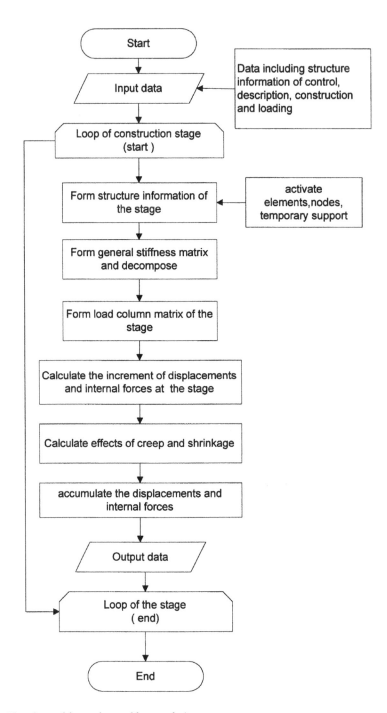

FIGURE 7.6 Flowchart of forward assemblage analysis.

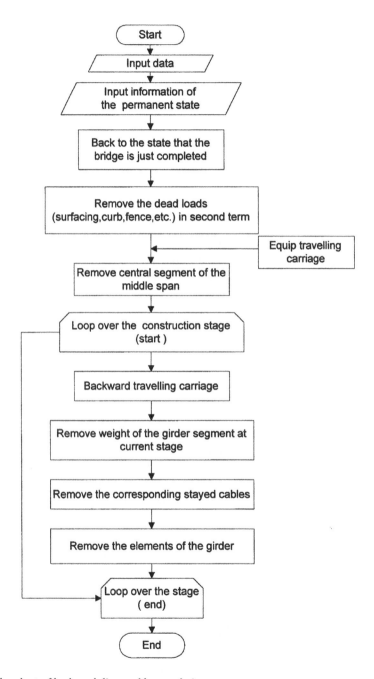

FIGURE 7.7 Flowchart of backward disassemblage analysis.

As mentioned above, for the erecting method using conventional form traveler cast-in-place or precast concrete segments, the crane or the form traveler may be modeled as external loads. Thus the carriage moving is equivalent to a change of the loading position. However, for an erection method utilizing cable-supported traveling carriage the cable stays first work as supports of the carriage and later, after curing is finished, the cable stays are connected with the girder permanently. In backward disassemblage analysis, the form traveler moving must be related to a change of the structural system.

The backward analysis procedure can establish the necessary data for the erection at each stage such as the elevations of deck, the cable forces, the deformations of structure, and the stresses at critical sections of deck and pylon.

One of the disadvantages of backward analysis is that creep effects are not able to be estimated; therefore, forward and backward simulations should be used alternately to determine the initial tension and the length of stay cables.

7.5 Construction Control

7.5.1 Objectives and Control Means

Obviously, the objective of construction control is to build a bridge that achieves the design aim with acceptable error. During the construction of a cable-stayed bridge, some discrepancies may occur between the actual state and the state of design expectation [14,15]. The discrepancies may arise from elevation error in laying forms, errors in stressing cable stays by jacks, errors of the first part of the dead load, i.e., the self-weights of the girder segments, and the second part of the dead load, i.e., the self-weights of the surfacing, curbs and fencing, etc. On the other hand, a system error may occur in measuring the deflection of the girder and the pylons. It is impossible to eliminate all the errors. Actually, there are two basic requirements for the completed structure [12]: (1) the geometric profile matches the designed shape well and (2) the internal forces are within the designed envelope values; specifically, the bending moments of the girder and the pylons are small and evenly distributed.

Since the internal forces of the girder and the pylons are closely related to the cable forces, the basic method of construction control is to adjust the girder elevation and the cable forces. If the error of the girder elevation deviated from the design value is small, such error can be reduced or eliminated by adjusting the elevation of the segment without inducing an angle between two adjacent deck segments. In this way we only change the geometric position of the girder without changing the internal force state of the structure. When the errors are not small, it is necessary to adjust the cable forces. In this case, both the geometric position change and the changes of internal forces occur in the structure.

Nevertheless, cable force adjustment are not preferable because they may take a lot of time and money. The general exercise at each stage is to find out the correct length of the cables and set the elevation of the segment appropriately. Cable tensioning is performed for the new stays only. Generally, a comprehensive adjustment of all the cables is only applied before connecting the two cantilever ends [21]. In case a group of cables needs to be adjusted, careful planning for the adjustment based on a detailed analysis is absolutely necessary.

7.5.2 Construction Control System

To guarantee structural safety and to reach the design aim, a monitoring and controlling system is important [13,15,19]. A typical construction controlling system consists of four subsystems: measuring subsystem, error and sensitivity analysis subsystem, control/prediction subsystem, and new design value calculation subsystem. An example of a construction control system for a PC cable-stayed bridge [18,20] is shown in Figure 7.8.

1. Measuring subsystem — The measuring items mainly include the elevation/deflection of the girder, the cable forces, the horizontal displacement of the pylon(s), the stresses of sections in the girder and the pylon(s), the modulus of elasticity and mass density of concrete, the creep and shrinkage of concrete, and the temperature/temperature gradient in the structure.

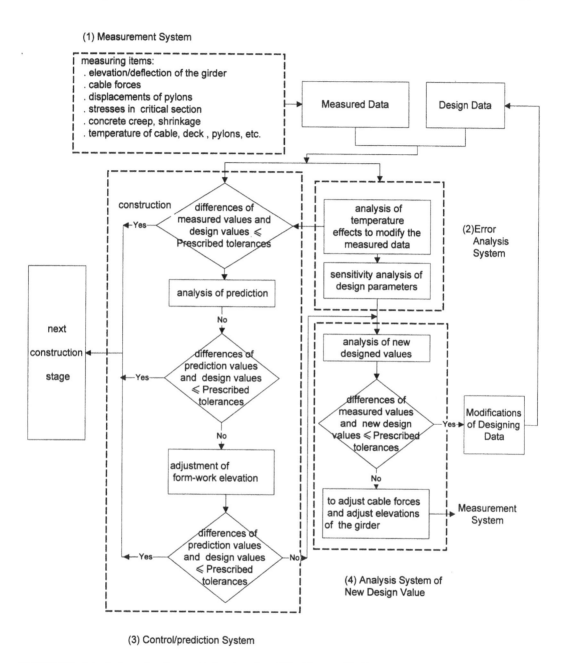

FIGURE 7.8 A typical construction control system for a PC cable-stayed bridge.

2. Error and sensitivity analysis subsystem — In this subsystem, the temperature effects are first determined and removed. Then the sensitivity of structural parameter such as the elasticity modulus of concrete, self-weight, stiffness of the girder segment or pylon, etc., are analyzed. Through the analysis, the causes of errors can be found so that the corresponding adjustment steps are utilized.

3. Control/prediction subsystem — Compare the measuring values with those of the design expectation; if the differences are lower than the prescribed limits, then go to the next stage. And the elevation is determined appropriately. Otherwise, it is necessary to find out the reasons, then to eliminate or reduce those errors through proper measures. The magnitude of cable tension adjustment can be determined by a linear programming model.

4. New design value calculation subsystem — Since the structural parameters for the completed part have deviated from the designed values, the design expectation must be updated with the changed state of the structure. And the sequential construction follows new design values so that the final state of structure can be achieved optimally.

7.6 An Engineering Example

The general view of a PC cable-stayed bridge is shown in Figure 7.1. The cable-stayed portion of the bridge has a total length of 702 m. The main span between towers is 380 m and the side anchor span is 161 m. The side anchor spans consist of two spans of 90 and 71 m with an auxiliary pier. The main girder is composed with two edge girders and a deck plate. The edge girders are laterally stiffened by a T-shaped PC girder with 6-m spacing. The edge girder is a solid section whose height is 2.2 m and width varies from 2.6 m at intersection of girder and pylon to 2.0 m at middle span. The deck plate is 28 cm thick. The width of the deck is 37.80 m out-to-out with eight traffic lanes. Spatial 244 stay cables are arranged in a semifan configuration. The pylon is shaped like a diamond with an extension mast (see Figure 7.1). All the cable stays are anchored in the mast part of the pylon. The stay cables are attached to the edge girders at 6 m spacing.

At the side anchor span an auxiliary pier is arranged to increase the stiffness of the bridge. And an anchorage segment of deck is set up to balance the lifting forces from anchorage cables.

7.6.1 Construction Process

The bridge deck structure is erected by the balance cantilevering method utilizing cable-supported form travelers. The construction process is briefly described as follows.

- Build the towers.
- Cast in place the first segment on timbering support.
- Erect the No. 1 cable and stress to its final length.
- Hoist the traveling carriages and positioning.
- Erect the girder segments one by one on the two sides of the pylons.
- Connect the cantilever ends of the side span with the anchorage parts.
- Continue to erect the girder segments in the center span.
- Connect the cantilever ends of the center span.
- Remove traveling carriages and temporary supports.
- Connect the girder with the auxiliary piers.
- Cast pavement and set up fence, etc.

A typical erection stage of one segment is described as follows:

- Move the traveling carriage forward and set up the form at proper levels.
- Erect and partially stress the stay cables attached to the traveler.
- Place reinforcement, post-tensioning bars and couple the stressed bars with those of the previously completed deck segment.
- Cast in place the deck concrete.
- Stress the stay cables to adjust the girder segments to proper levels.
- Cure deck panel and stress the longitudinal and lateral bars and strands.
- Loosen the connection between the stay cable and traveling carriage.
- Stress cable stays to the required value.

The above erection steps are repeated until the bridge is closed at the middle span.

TABLE 7.1 Predicted Initial Cable Forces (kN)

No.	S1	S2	S3	S4	S5	S6	S7	S8	S9	S10
NCS	7380	3733	4928	5130	5157	5373	5560	5774	5969	6125
CS	7745	3854	5101	5331	5348	5549	5694	5873	6050	6185

No.	S11	S12	S13	S14	S15	S16	S17	S18	S19	S20
NCS	6320	6870	7193	7283	7326	7519	7317	7478	7766	8035
CS	6367	6908	7209	7331	7467	7639	7429	7580	7856	8114

No.	S21	S22	S23	S24	S25	S26	S27	S28	S29	S30
NCS	8366	8169	7693	8207	9579	9649	9278	9197	9401	12820
CS	8442	8246	7796	8354	9778	9791	9509	9296	9602	12930

No.	M1	M2	M3	M4	M5	M6	M7	M8	M9	M10
NCS	7334	3662	4782	4970	5035	5426	5479	5628	5778	6108
CS	7097	3433	4591	4877	5006	5477	5567	5736	5904	6231

No.	M11	M12	M13	M14	M15	M16	M17	M18	M19	M20
NCS	6139	6572	6627	6918	6973	7353	7540	7742	7882	7978
CS	6263	6706	6756	7031	7088	7422	7624	7813	7942	8060

No.	M21	M22	M23	M24	M25	M26	M27	M28	M29	M30
NCS	8384	8479	8617	8833	9175	9359	9394	9480	9641	13440
CS	8452	8561	8694	8931	9212	9413	9459	9570	9716	13570

No.: Cable number; M: middle span; S: side span; CS: with the effects of creep and shrinkage of concrete; NCS: without the effects of creep and shrinkage of concrete.

7.6.2 Construction Simulation

The above construction procedure can be simulated stage by stage as illustrated in Section 7.4.2. Since creep and shrinkage occur and the second part of the dead weight will be loaded on the bridge girder after completion of the structure, a downward displacement is induced. Therefore, as the erection is just finished the elevation of the girder profile should be set higher than that of the design profile and the pylons should be leaning toward the side spans. In this example, the maximum value which is set higher than the designed profile in the middle of the bridge is about 35.0 cm, while the displacement of pylon top leaning to anchorage span is about 9.0 cm. The initial cable forces are listed in Table 7.1 to show the effects of creep. As can be seen, considering the long-term effects of concrete creep, the initial cable forces are a little greater than those without including the time-dependent effects.

7.6.3 Construction Control System

In the construction practice of this PC cable-stayed bridge, a construction control system is employed to control the cable forces and the elevation of the girder. Before starting concrete casting the reactions of the cable-supported form traveler are measured by strain gauge equipment. Thus the weights of the four travelers used in this bridge are known.

At each stage the mass density of concrete and the elasticity modulus of concrete are tested in the laboratory *in situ*. The calculation of construction is carried out with the measured parameters. In several sections of the deck and pylon, strain gauges are embedded to measure the strains of the structure during the whole construction period, thus the stress of the structure can be monitored.

In this example the main flowchart of the construction control system for a typical erecting segment is shown in Figure 7.9.

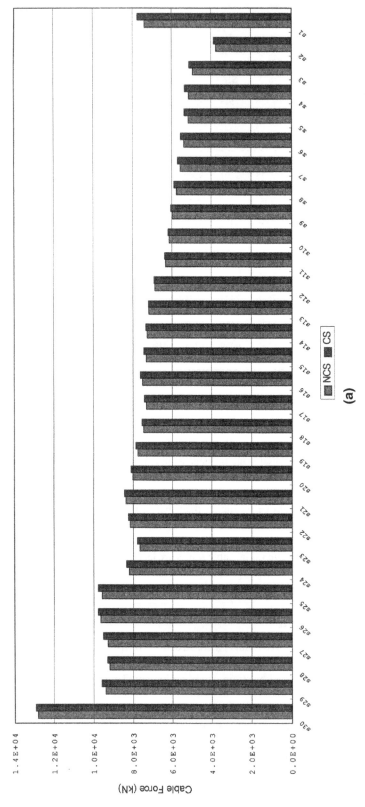

FIGURE 7.9 (a) Initial cable forces in side span determined by simulation analysis. (b) Initial cable forces in middle span determined by simulation analysis.

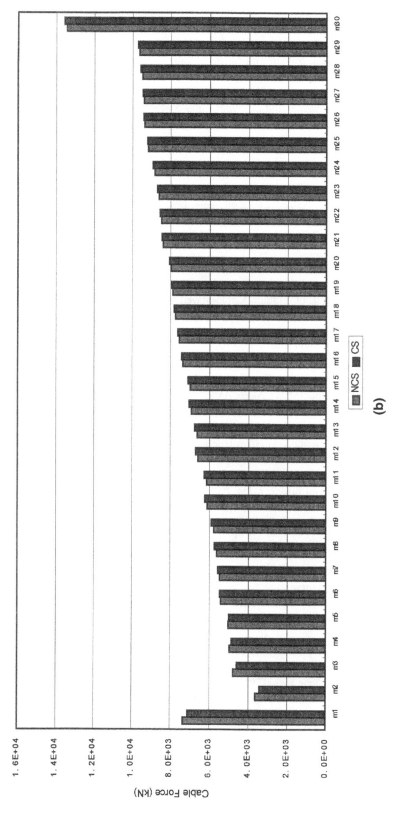

FIGURE 7.9 (continued)

References

1. Tang, M.C. The 40-year evolution of cable-stayed bridges, in *International Symposium on Cable-Stayed Bridges,* Lin Yuanpei et al., Eds., Shanghai, 1994, 30–11.
2. Leonhardt, F. and Zellner, W., Past, present and future of cable-stayed bridges, in *Cable-Stayed Bridges, Recent Developments and Their Future,* M. Ito et al., Eds., Elsevier Science Publishers, New York, 1991.
3. Podolny, W. and Scalmi, J., *Construction and Design of Cable-Stayed Bridges,* John Wiley & Sons, New York, 1983.
4. Walther, R., Houriet, B., Lsler, W., and Moia, P., *Cable-Stayed Bridges,* Thomas Telford, London, 1988.
5. Gimsing, N. J., *Cable Supported Bridges, Concept and Design,* John Wiley & Sons, New York, 1983.
6. Kasuga, A., Arai, H., Breen, J. E., and Furukawa, K., Optimum cable-force adjustment in concrete cable-stayed bridges, *J. Struct. Eng.,* ASCE, 121(4), 685–694, 1995.
7. Ma, W. T., Cable Force Adjustment and Construction Control of PC Cable-Stayed Bridges, Ph.D. dissertation of Department of Civil Engineering, South China University of Technology, 1997 [in Chinese].
8. Wang, X. W. et al., A study of determination of cable tension under dead loads, *Bridge Constr.,* 4, 1–5, 1996 [in Chinese].
9. Yan, D. H. et al., Simulation analysis of Tongling Cable-Stayed Bridge for construction control, in *National Symposium on Highway Bridge,* Dai Jing, Ed., Beijing Renming Jiaotong Press, Guangzhou, 347–355, 1995 [in Chinese].
10. Zhou, L. X. et al., Prestressed Concrete Cable-Stayed Bridges, Beijing Renming Jiaotong Press, 1989 [in Chinese].
11. Xiao, R. C., Ling, P. Application of computational structural mechanics in construction design and control of bridge structures, *Comput. Struct. Mech. Appl.,* 10(1) 92–98, 1993 [in Chinese].
12. Fang, Z. and Liu, G. D., A Study of Construction Control System of Cable-Stayed Bridges, Research Report of Department of Civil Engineering, Hunan University, 1995 [in Chinese].
13. Chen, D. W., Xiang, H. F., and Zheng, X. G., Construction control of PC cable-stayed bridge, *J. Civil Eng.,* 26(1) 1–11, 1993 [in Chinese].
14. Yoshimura, M., Ueki, Y., and Imai, Y., Design and construction of a prestressed concrete cable-stayed bridge: the Tsukuhara Ohashi Bridge, *J. Jpn. Prestressed Concrete Eng. Assoc.,* Tokyo, Japan, 29(1) 1987 [in Japanese].
15. Fujisawa, N. and Tomo, H., Computer-aided cable adjustment of cable-stayed bridges, *IABSE Proc.,* P-92/85, 1985.
16. Furukawa, K., Inoue, K., Nakkkayama, H., and Ishido, K., Studies on the management system of cable-stayed bridges under construction using multi-objective programming method. *Proc. JSCE,* Tokyo, Japan, 374(6), 1986 [in Japanese].
17. Furuta, H. et al., Application on fuzzy mathematical programming to cable tension adjustment of cable-stayed bridges, in *International Symposium on Cable-Stayed Bridges,* Lin Yuanpei et al., Eds., Shanghai, 1994, 584–595.
18. Takuwa, I. et al., Prestressed concrete cable-stayed bridge constructed on an expressway — the Tomei Ashigra Bridge, in *Cable-Stayed Bridges, Recent Developments and Their Future.* M. Ito et al., Eds., Elsevier Science Publishers, New York, 1991.
19. Yasuhiro, K. et al., Construction of Tokachi Ohashi Bridge Superstructure (PC cable-stayed bridge), *Bridge Found.,* 1, 7–15, 1995 [in Japanese].
20. Hidemi, O. et al., Construction of Ikara Bridge superstructure (PC cable-stayed bridge), *Bridge Found.,* No. 11, 7–14, 1995 [in Japanese].
21. Fushimi, T., et al., Erection of the Tsurumi Fairway Bridge superstructure, *Bridge Found.,* 10, 2–10, 1994 [in Japanese].

Index

Printed and bound by CPI Group (UK) Ltd, Croydon, CR0 4YY

23/10/2024

01778248-0010